METHODS *IN* YEAST GENETICS

A Cold Spring Harbor Laboratory Course Manual

1997 Edition

Alison Adams
University of Arizona

Daniel E. Gottschling
Fred Hutchinson Cancer Research Center

Chris A. Kaiser
Massachusetts Institute of Technology

Tim Stearns
Stanford University

COLD SPRING HARBOR LABORATORY PRESS

METHODS *IN* YEAST GENETICS

A Cold Spring Harbor Laboratory Course Manual

1997 Edition

Production Editor/Project Coordinator: Maryliz M. Dickerson

ISBN 0-87969-508-0
Library of Congress Catalog Card Number: 97-77211

All suppliers mentioned in this manual can be found in the BioSupplyNet Source Book and on the Web site at: http://www.biosupplynet.com.

If a copy of BioSupplyNet Source Book was not included with this manual, a free copy can be ordered by using any of the following methods:

- Complete the Free Source Book Request Form found at the Web site at: http://www.biosupplynet.com
- E-mail a request to info@biosupplynet.com
- Fax a request to 516-349-5598

All Cold Spring Harbor Laboratory Press publications may be ordered directly from Cold Spring Harbor Laboratory Press, 10 Skyline Drive, Plainview, New York 11803-2500. Phone: 1-800-843-4388 in Continental U.S. and Canada. All other locations: (516) 349-1930. FAX: (516) 349-1946.

Contents

Preface

This laboratory course manual incorporates significant portions of the manuals used in previous Cold Spring Harbor Yeast Genetics Courses. Although most of the experiments have now been revised and several new techniques have been added, the basic emphases of this course are the same as those that were taught in the Cold Spring Harbor Yeast Genetics Course for 27 years. We are indebted to our predecessors, Fred Sherman, Gerry Fink, Jim Hicks, Mark Rose, Fred Winston, Phil Hieter, Susan Michaelis, and Aaron Mitchell for their teaching and for making this course an important part of the yeast community. We also thank Mike Cherry for his invaluable assistance with the genetic and physical maps.

A. Adams
D.E. Gottschling
C.A. Kaiser
T. Stearns

Introduction

Genetic investigations of yeast were essentially initiated by Winge and his co-workers in the mid-1930s. Approximately ten years later, Lindegren and his colleagues also began extensive studies. These two groups are responsible for uncovering the general principles and much of the basic methodology of yeast genetics. Today, yeast is widely recognized as an ideal eukaryotic microorganism for biochemical and genetic studies. Although yeasts have greater genetic complexity than bacteria, they still share many of the technical advantages that permitted rapid progress in the molecular genetics of prokaryotes and their viruses. Some of the properties that make yeast particularly suitable for genetic studies include the existence of both stable haploid and diploid cells, rapid growth, clonability, the ease of replica plating, mutant isolation, and ability to isolate each haploid product of meiosis by microdissection of a tetrad ascus. Yeast has been successfully employed for the study of all areas of genetics, such as mutagenesis, recombination, chromosome segregation, gene action and regulation, as well as aspects distinct to eukaryotic systems, such as mitochondrial genetics.

DNA transformation has made yeast particularly accessible to gene cloning and genetic engineering techniques. Structural genes corresponding to virtually any genetic trait can be identified by complementation from plasmid libraries. DNA is introduced into yeast cells either as replicating molecules or by integration into the genome. In contrast to most other organisms, integrative recombination of transforming DNA in yeast proceeds primarily via homologous recombination. This permits efficient targeted integration of DNA sequences into the genome. Homologous recombination coupled with high levels of gene conversion has led to the development of techniques for the direct replacement of normal chromosomal loci with genetically engineered DNA sequences. This ease of performing direct gene replacement is unique among eukaryotic organisms and has been extensively exploited in every aspect of yeast genetics, cell biology, physiology, and biochemistry. Many of these modern genetic techniques are reviewed in Volume 194 of *Methods in Enzymology* (Guthrie and Fink 1991) and by Rose (1995).

The latest great achievement of modern genetics has been determining the complete DNA sequence of the *Saccharomyces cerevisiae* genome. The assembly of this sequence into a public sequence database has made yeast the premier organism for detailed analysis of eukaryotic cellular function and genome organization. Several Internet web sites provide convenient and organized access to the sequence database. Two of the more popular sites are the Saccharomyces Genome Database (SGD) and the Munich Information Center for Protein Sequences (MIPS). Each of these sources augments the sequence database by providing powerful search functions for analyzing the genome, as well as helpful links to relational databases, such as literature cross-references or protein structure of yeast gene products. In addition, the Yeast Protein Database (YPD) has each yeast protein and predicted open reading frame compiled with information about its structure and the phenotypes associated with mutations in its gene.

One of the most useful collections of information on *Saccharomyces cerevisiae* biology are found in two companion series of reviews entitled *The Molecular Biology of the Yeast* Saccharomyces (published in 1981 and 1982) and *The Molecular and Cellular Biology of the Yeast* Saccharomyces (published in 1991–1997). While a few chapters in the later series update progress made since the first series was published, each volume has detailed reviews that synthesize vast literatures of yeast biology that are not available anywhere else. All five of these books are essential for the library of a yeast biologist. *The Early Days of Yeast Genetics* is also a book to consider reading, for it provides a historical perspective on why yeast became a model organism and traces the development of the yeast community's influence on modern genetics and eukaryotic molecular biology.

This course will focus exclusively on the baker's yeast, *Saccharomyces cerevisiae*. After completing the course, you should be able to carry out all of the techniques commonly employed by yeast geneticists and be able to follow the literature with greater ease. Except for the dissection of asci, most of the methods do not differ significantly from the methods employed with other microorganisms, and the skills should be rapidly acquired with little practice. Experiments will be conducted in pairs, and whenever possible, an investigator more familiar with microbiological techniques will be assigned to a less-experienced partner. Since most of the experiments will be initiated at the outset of the course, it is advisable to read the entire manual thoroughly. Please note that some experiments span many days due to extended periods of incubation.

We wish to emphasize that some of the procedures in this manual have been condensed in order to save time and are not necessarily standard for research purposes. For example, mutants are usually purified by subcloning the initial isolates. Also, some of the techniques may not be directly applicable to research problems, but they have been included to illustrate general principles and methods.

HIGHLY RECOMMENDED BOOKS

Broach, J.R., E.W. Jones, and J.R. Pringle. 1991. *The molecular and cellular biology of the yeast* Saccharomyces, vol. 1, *Genome dynamics, protein synthesis, and energetics.* Cold Spring Harbor Laboratory Press, Cold Spring Harbor, New York.

Guthrie, C. and G.R. Fink, eds. 1991. *Methods in Enzymology*, vol. 194, *Guide to Yeast Genetics and Molecular Biology.* Academic Press, New York.

Jones, E.W., J.R. Pringle, and J.R. Broach. 1992. *The molecular and cellular biology of the yeast* Saccharomyces, vol. 2, *Gene expression.* Cold Spring Harbor Laboratory Press, Cold Spring Harbor, New York.

Pringle, J.R., J.R. Broach, and E.W. Jones. 1997. *The molecular and cellular biology of the yeast* Saccharomyces, vol. 3, *Cell cycle and cell biology.* Cold Spring Harbor Laboratory Press, Cold Spring Harbor, New York.

Strathern, J.N., E.W. Jones, and J.R. Broach, eds. 1981. *The molecular biology of the yeast* Saccharomyces: *Life cycle and inheritance.* Cold Spring Harbor Laboratory, Cold Spring Harbor, New York.

———. 1982. *The molecular biology of the yeast* Saccharomyces: *Metabolism and gene expression.* Cold Spring Harbor Laboratory, Cold Spring Harbor, New York.

OTHER BOOKS ABOUT YEAST

Fincham, J.R.S., P.R. Day, and A. Radford. 1979. *Fungal genetics.* University of California Press, Berkeley and Los Angeles.

Rose, M.D. 1995. Modern and post-modern genetics in Saccharomyces cerevisiae. In *The yeasts* (ed. A.E. Wheals et al.), 2nd ed., vol. 6. Academic Press, New York.

Hall, M.N., and P. Linder. 1993. *The early days of yeast genetics.* Cold Spring Harbor Laboratory Press, Cold Spring Harbor, New York.

RECOMMENDED YEAST WEB SITES

Munich Information Center for Protein Sequences
http://speedy.mips.biochem.mpg.de/mips/yeast/

Saccharomyces Genome Database
http://genome-www.stanford.edu/Saccharomyces/

Yeast Protein Database
http://www.proteome.com/YPDhome.html

Genetic Nomenclature

CHROMOSOMAL GENES

Early recommendations for the nomenclature and conventions used in yeast genetics have been summarized by Sherman and Lawrence (1974) and Sherman (1981). Whenever possible, gene symbols are consistent with the proposals of Demerec et al. (1966) and are designated by three italicized letters (e.g., *arg*). Contrary to the proposals of Demerec et al. (1966), the genetic locus is identified by a number (not a letter) following the gene symbol (e.g., *arg2*). Dominant alleles are denoted by using uppercase italics for all three letters of the gene symbol (e.g., *ARG2*). Lowercase letters symbolize the recessive allele (e.g., the auxotroph *arg2*). Wild-type genes are designated with a superscript plus sign ($sup6^+$ or $ARG2^+$). Alleles are designated with a number separated from the locus number by a hyphen (e.g., *arg2-14*). Locus numbers are consistent with the original assignments; however, allele numbers may be specific to a particular laboratory.

Phenotypic designations are sometimes denoted by cognate symbols in Roman type followed by a superscript plus or minus sign. For example, the independence from and requirement for arginine can be denoted by Arg^+ and Arg^-, respectively.

The following examples illustrate the conventions used in the genetic nomenclature for *S. cerevisiae:*

ARG2	A locus or dominant allele
arg2	A locus or recessive allele that produces a requirement for arginine as the phenotype
$ARG2^+$	The wild-type allele of this gene
arg2-9	A specific allele or mutation at the *ARG2* locus
Arg^+	A strain that does not require arginine
Arg^-	A strain that requires arginine
Arg2p	Designation for the protein product of the *ARG2* gene

There are a number of exceptions to these general rules. Gene clusters, complementation groups within a gene, or domains within a gene that have different properties can be designated by capital letters following the locus number (e.g., *his4A, his4B*).

The extensive use in yeast of recombinant DNA techniques has introduced a nomenclature that pertains to gene insertions, gene fusions, and plasmids:

ARG2::LEU2	An insertion of the *LEU2* gene at the *ARG2* locus where the insertion does not disrupt *ARG2* function
arg2::LEU2	An insertion of the *LEU2* gene at the *ARG2* locus where the insertion disrupts *ARG2* function
arg2-101::LEU2	An insertion of the *LEU2* gene at the *ARG2* locus where the insertion disrupts *ARG2* function and the disruption allele is specified
cyc1–arg2	A gene fusion between the *CYC1* gene and *ARG2* where neither gene is functional
P_{cyc1}–*ARG2*	A gene fusion between the *CYC1* gene promoter and *ARG2* where the *ARG2* gene is functional
[YCp–*ARG2*]	A centromere plasmid carrying a functional *ARG2* locus
[pCK101]	Designation for a specific plasmid whose structure is given elsewhere

Although superscripts should be avoided, it is sometimes expedient to distinguish genes conferring resistance or sensitivity by a superscript R or S, respectively. For example, the genes controlling resistance to canavanine sulfate (*can1*) and copper sulfate (*CUP1*) and their sensitive alleles can be denoted, respectively, as *can*^{R}*1*, *CUP*^{R}*1*, *CAN*^{S}*1*, and *cup*^{S}*1*.

Wild-type and mutant alleles of the mating-type and related loci do not follow the standard rules. The two wild-type alleles of the mating-type locus are designated *MAT***a** and *MAT*α. The two complementation groups of the *MAT*α locus are denoted *MAT*α*1* and *MAT*α*2*. Mutations of the *MAT* genes are denoted, e.g., *mat***a**-*1*, *mat*α*1-1*. The wild-type homothallic alleles at the *HMR* and *HML* loci are denoted *HMR***a**, *HMR*α, *HML***a**, and *HML*α. Mutations at these loci are denoted, e.g., *hmr***a**-*1*, *hml*α-*1*. The mating phenotypes of *MAT***a** and *MAT*α cells are denoted simply **a** and α, respectively.

Dominant and recessive suppressors should be denoted, respectively, by three uppercase or three lowercase letters followed by a locus designation (e.g., *SUP4*, *SUF1*, *sup35*, *suf11*). In some instances, UAA suppressors and UAG suppressors are further designated o and a, respectively, following the locus. For example, *SUP4*-o refers to suppressors of the *SUP4* locus that insert tyrosine residues at UAA sites; *SUP4*-a refers to suppressors of the same *SUP4* locus that insert tyrosine residues at UAG sites. The corresponding wild-type locus coding for the normal tyrosine tRNA and lacking suppressor activity can be referred to as *sup4*$^{+}$. Thus, the nomenclature describing suppressor and wild-type alleles in yeast is unrelated to the bacterial nomenclature. For example, an ochre *E. coli* suppressor that inserts tyrosine residues at both UAA and UAG sites is denoted as su_4^+, and the wild-type locus coding for the normal tyrosine tRNA and lacking suppressor activity can be referred to as Su_4, su_4^-, or *supC*.

For most structural genes that code for proteins, the functional wild-type allele is usually dominant to the mutant form of a gene. In yeast, the convention for dominant genes utilizes italic symbols such as *HIS4* and *LEU2*. Because the sites of recessive mutations are usually used for genetic mapping, published chromosome maps usually contain the mutant form of the gene. For example, chromosome III contains *his4* and *leu2*, whereas chromosome IX contains *SUP22* and *FLD1*. Because capital letters are used to represent dominant wild-type genes that control the same character (e.g., *SUC1*, *SUC2*) and because the dominant forms are used in genetic mapping, such chromosomal loci are denoted in capital letters on genetic maps. In addition, capital letters are used to designate certain DNA segments whose locations have been determined by a combination of recombinant DNA techniques and classical mapping procedures (e.g., *RDN1*, the segment encoding ribosomal RNA).

NON-MENDELIAN DETERMINANTS

Where necessary, non-Mendelian genotypes can be distinguished from chromosomal genotypes by enclosure in brackets. Whenever applicable, it is advisable to employ the above rules for designating non-Mendelian genes and to avoid the use of Greek letters. However, when referring to an entire non-Mendelian element, it is best to either retain the original symbols [ρ^+], [ρ^-], [ψ^+], and [ψ^-] or use their transliteration, [*rho*$^+$], [*rho*$^-$], [*PSI*$^+$], and [*psi*$^-$], respectively. Detailed designations for mitochondrial mutants have been presented by Dujon (1981) and Grivell (1984, 1990) and for killer strains by Wickner (1981). Unlike the other non-Mendelian determinants, [*PSI*$^+$] and [*URE3*] are not based on different states of a nucleic acid; rather, the [*PSI*$^+$] and [*URE3*] traits result from heritable conformational states of proteins. The unusual behavior of these traits can be explained by the prion hypothesis, which has been used to explain infectious neurodegenerative diseases in mammals such as scrapie (Lindquist 1997). [*PSI*$^+$] corresponds to a heritable conformational state of the translation termination factor Sup35p, and [*URE3*] corresponds to a heritable inactive state of *URE2*, a gene whose product is involved in nitrogen regulation. The known non-Mendelian determinants in yeast are listed in Table 1.

Table 1 Non-Mendelian Determinants of Yeast

Wild type	Mutant variant	Element	Mutant trait
[ρ^+]	[ρ^-]	mitochondrial DNA	respiration deficiency
[*KIL*-k_1]	[*KIL*-o]	RNA plasmid	sensitive to killer toxin
[*cir*$^+$]	[*cir*^{o}]	2μ plasmid	none
[*psi*$^-$]	[*PSI*$^+$]	prion form of Sup35p	enhanced suppression of nonsense codons
[*ure3*$^-$]	[*URE3*]	prion form of Ure2p	unregulated ureidosuccinate uptake

GENETIC BACKGROUNDS

The genetic background from which a *S. cerevisiae* strain is derived is an often hidden aspect of the genotype that should be taken into account when designing experiments. Most strains used in modern genetic studies come from one of a small set of genetic backgrounds, including S288C, X2180, A364A, W303, Σ1278b, AB972, SK1, and FL100. The genealogies of some of these backgrounds have recently been reconstructed from records of crosses that were carried out in the 1940s between wild yeasts and brewing strains (Mortimer and Johnston 1986). This analysis shows that although most backgrounds share a common ancestry, a significant degree of genetic heterogeneity has been introduced by outcrossing. In practice, crosses between distantly related strains often give inviable combinations of alleles leading to many inviable spores, whereas crosses between strains from the same background usually give >95% viable spores. The S288C and A364A genetic backgrounds have similar genealogies and in crosses give a high frequency of spore viability, but an analysis of genomic sequences from these strains reveals an average of 3.4 nucleotide sequences differences per kilobase of genomic DNA. Thus, even apparently closely related strains can differ at a very large number of sites.

Allelic differences between strain backgrounds can seriously influence the outcome of many different kinds of experiments, and it is best to avoid genetic heterogeneity as much as possible by using a single genetic background. At the beginning of a new mutant hunt, it is worth considering which strain background to use—usually the background used by most investigators in the same field is the best choice. S288C is probably the most commonly used background; however, other backgrounds offer distinct advantages for particular types of experiments. For example, Σ1278b will form pseudohyphae whereas S288C will not, and SK1 (the strain we use for tetrad analysis in Experiment III) sporulates much more rapidly than S288C. It is often necessary to move a desired mutation from one background into another. Ideally, this can be done using recombinant plasmids and the methods for gene replacement described in Experiment VII. Mutations that have not been cloned can be moved by backcrossing to the desired strain background (usually successive backcrosses are carried out until a clear 2:2 pattern of segregation for the desired trait has been achieved).

REFERENCES

Demerec, M., E.A. Adelberg, A.J. Clark, and P.E. Hartman. 1966. A proposal for a uniform nomenclature in bacterial genetics. *Genetics* **54:** 61–76.

Dujon, B. 1981. Mitochondrial genetics and functions. In *The molecular biology of the yeast* Saccharomyces: *Life cycle and inheritance* (ed. J.N. Strathern et al.), pp. 505–635. Cold Spring Harbor Laboratory, Cold Spring Harbor, New York.

Grivell, L.A. 1984. Restriction and genetic maps of yeast mitochondrial DNA. In *Genetic maps*, 3rd edition (ed. S.J. O'Brien), pp. 234–247. Cold Spring Harbor Laboratory, Cold Spring Harbor, New York.

———. 1990. Mitochondrial DNA in the yeast *Saccharomyces cerevisiae.* In *Genetic maps*, 5th edition (ed. S.J. O'Brien), pp. 3.50–3.57. Cold Spring Harbor Laboratory Press, Cold Spring Harbor, New York.

Lindquist, S. 1997. Mad cows meet psi-chotic yeast: The expansion of the prion hypothesis. *Cell* **89:** 495–498.

Mortimer, R.K. and J.R. Johnston. 1986. Genealogy of principal strains of the yeast genetic stock center. *Genetics* **113:** 35–43.

Sherman, F. 1981. Genetic nomenclature. In *The molecular biology of the yeast* Saccharomyces: *Life cycle and inheritance* (ed. J.N. Strathern et al.), pp. 639–640. Cold Spring Harbor Laboratory, Cold Spring Harbor, New York.

Sherman, F. and C.W. Lawrence. 1974. *Saccharomyces.* In *Handbook of genetics: Bacteria, bacteriophages, and fungi* (ed. R.C. King), vol. 1, pp. 359–393. Plenum Press, New York.

Wickner, R.B. 1981. Killer systems in *Saccharomyces cerevisiae*. In *The molecular biology of the yeast* Saccharomyces: *Life cycle and inheritance* (ed. J.N. Strathern et al.), pp. 415–444. Cold Spring Harbor Laboratory, Cold Spring Harbor, New York.

EXPERIMENT I

Looking at Yeast Cells

Yeast cells are approximately 5 μm in diameter, and many of their important features can be seen in the light microscope. It is good laboratory practice to routinely examine cultures under phase microscopy for indications of the physiological state of the cells, and for evidence of contamination. Much of modern yeast cell biological work involves more sophisticated microscopic examination of yeast cells stained with protein-specific antibodies, or with fluorescent dyes that specifically associate with certain organelles. This experiment will provide examples of the standard types of light microscopy that are used in the examination of yeast cells.

EXAMINATION OF GROWING CULTURES

Growth Properties

Saccharomyces cerevisiae cells grow by budding. A cell that gives rise to a bud is called a mother cell, and the bud is sometimes referred to as the daughter cell. A new bud emerges from a mother cell close to the beginning of the cell cycle and continues to grow throughout the cell cycle until it separates from the mother cell at the end of the cell cycle. Because all of the growth of a yeast cell is concentrated in the bud, and because this growth is essentially continuous throughout the cell cycle, the size of the bud gives an approximate indication of the position of a given cell in the cell cycle. An exponentially growing culture of yeast cells has approximately one-third unbudded cells, one-third cells with a small bud, and one-third cells with a large bud. When cells in a growing culture use up the available nutrients, they stop growing by arresting in the cell cycle as unbudded cells. Thus, a simple way of determining the growth state of a culture is to determine the frequency of budded cells in the microscope. Note that for some strains, the mother and daughter cells remain stuck together even though they have completed cytokinesis. In these cases, it is necessary to vortex or sonicate the culture to separate cells prior to microscopy. Many kinds of mutants also arrest in the cell cycle in a way that is diagnostic of their phenotype. For example, cells in which there is a defect in the mitotic spindle arrest as large budded cells, a point in the cycle that would normally correspond to mitosis. It is important to note that the arrest point, or terminal phenotype, of mutant cells can be morphologically distinct from any cell type seen in a normal culture. In the mitotic mutant above, the mother and daughter cells continue to grow at the arrest point until both are much larger than normal yeast cells.

Haploids vs. Diploids

Haploid and diploid yeast cells are morphologically similar but differ in several important ways. First, diploid cells are larger than haploid cells. Cytoplasmic volume increases with ploidy, and the diameter of a diploid cell is roughly 1.3 times that of a haploid cell. This difference can be readily seen when haploids and diploids are compared side by side. Because they are larger, diploid cells (or even tetraploids in some cases) are often used for fluorescence microscopy where the larger size helps in being able to resolve small cellular structures. Second, diploid cells tend to have a more elongated, or ovoid, shape than haploid cells, which are often almost round. Third, diploids and haploids have a different budding pattern. Yeast cells generally bud about 20 times before becoming senescent. Successive buds emerge from the surface of the mother cell in stereotyped patterns. Haploid cells bud in an axial pattern wherein each bud emerges adjacent to the site of the previous bud. Diploid cells bud in a polar pattern wherein successive buds can emerge from either end of the elongated mother cell. The history of a cell's budding pattern can be visualized by staining cells with Calcofluor, a fluorescent compound that binds to the rings of chitin that remain at old bud sites. These chitin rings are called bud scars, and we will use Calcofluor staining of haploid and diploid cells to visualize the axial and polar bud scar patterns.

Mating Cells

Yeast cells come in three mating types: *MAT***a** and *MAT*α; these two are able to mate with each other to yield a *MAT***a**/α. *MAT***a**/α cells cannot mate with cells of either mating type. Generally, *MAT***a** and *MAT*α strains will be haploids and MAT**a**/α strains will be diploid, although this is not always the case, and one should be careful to consider mating type independent of ploidy. The mating process between two cells begins with an exchange of pheromones that causes each of the cells to arrest in the cell cycle as unbudded cells and to induce the expression of proteins required for mating. The pheromone also causes the cells to make a projection of new cell surface specialized for cell fusion. This projection is usually oriented toward the mating partner. Cells with a mating projection are called ''shmoos'' because of their resemblance to an Al Capp cartoon character from the 1940s. The shmooing cells join at the tip of their projections, their cytoplasms fuse, and then their nuclei fuse to form a diploid *MAT***a**/α nucleus. The process of nuclear fusion is termed karyogamy. The newly formed diploid is termed a zygote and has a characteristic appearance that is particularly easy to identify when the first bud emerges. It is possible to isolate zygotes by micromanipulation, allowing for the isolation of diploid cells even in situations where there is no genetic selection for diploid formation. We will look at a population of mating cells to identify shmoos and zygotes.

Mitochondria

Yeast mitochondria contain a genome that encodes several proteins involved in oxidative phosphorylation, one ribosomal protein, and the rRNAs and tRNAs required for the mitochondrial translation apparatus. The vast majority of mitochondrial proteins are encoded by nuclear genes and imported into the mitochondria from the cytoplasm. Thus, mutations in either the nuclear genome or the mitochondrial genome can affect mitochondrial function. Yeast cells with mutations in the mitochondrial DNA cannot carry out oxidative phosphorylation, so must get all of their energy from fermentation. Such mutants grow more slowly than wild-type cells and are unable to grow on nonfermentable carbon sources such as lactate, glycerol or ethanol. The French scientists who first characterized mitochondrial mutants called this the "petite" phenotype. Petite strains form small milky-white colonies. This lack of colony pigmentation is most evident in an *ade2* background; formation of the red pigment that typifies *ade2* mutants requires oxidative phosphorylation. Diploid petite strains are also unable to sporulate, and it is wise to check a nonsporulating strain for the ability to grow on a nonfermentable carbon source before other potential causes of sporulation failure are examined. Petite mutants appear with high frequency in many common lab yeast strains—for some strains, as many as 10% of the cells in a culture are petite. Although the petite phenotype can be due to mutations in either the mitochondrial or nuclear genomes, the great majority of petites are due to mitochondrial DNA mutations. The mitochondrial genome is given the designation "rho"; wild-type strains are rho^+, strains with deleted versions of the mitochondrial genome (the most common type of mutation) are rho^-, and strains lacking the mitochondrial genome entirely are rho^o. A common misconception is that rho^o strains lack mitochondria altogether. Several essential reactions take place within the mitochondrial membrane, and even in rho^o strains a diminished mitochondrial structure can be seen in the electron microscope. In this experiment, we will use fluorescence microscopy to visualize the mitochondria and mitochondrial DNA in wild-type cells.

FLUORESCENCE MICROSCOPY

As yeast cells have been used increasingly for experiments in cell biology, methods for determining the intracellular localization of gene products have been developed, usually by adaptation of methods first used in animal cells. The association of gene products with known cellular structures has frequently led to important insights into the function of the genes involved. Moreover, examination of the morphologies of the nucleus and cytoskeleton gives precise information about the cell cycle stage of individual cells and has been used to stage cell division cycle mutants.

In this experiment, we will use three different methods of identifying structures or proteins in yeast cells using fluorescence microscopy. The first is immunofluorescence, in which fixed cells are incubated first with primary antibodies against the protein of interest, then with fluorescently-labeled secondary antibodies directed against the primary antibodies. This layering of primary and secondary antibodies increases the potential fluorescent signal. We will use immunofluorescence with anti-tubulin antibodies to visualize the microtubule cytoskeleton. This method has the advantage that it usually generates a strong fluorescence signal, but the disadvantage that cells must fixed prior to observation, and that the cell wall must be removed to allow access of the antibodies. Care must be taken to avoid artifacts due to the cell preparation.

The second fluorescence microscopy method makes use of small molecules that are fluorescent and either bind to specific proteins in the cell or are partitioned to certain organelles. We will use 4′6,-diamidino-2-phenylindole (DAPI) to stain DNA, 3,3′-dihexyloxacarbocyanine iodide ($DiOC_6$) to stain mitochondria, and rhodamine-phalloidin to stain the actin cytoskeleton. DAPI binds specifically to DNA and becomes more intensely fluorescent when bound. $DiOC_6$ is a fluorescent hydrophobic molecule that is specifically transported into mitochondria. Phalloidin is a toxin from the *Amanita phalloides* mushroom that binds to actin filaments; when labeled with rhodamine, it allows the fluorescent visualization of actin filaments. DAPI can be used on fixed or living cells, $DiOC_6$ must be used on living cells, and rhodamine-phalloidin is usually used on fixed cells.

The last and most recently developed method makes use of green fluorescent protein (GFP), a naturally fluorescent protein from the jellyfish *Aequoria victoria.* GFP is remarkable in that it retains its fluorescence when expressed in bacterial, fungal, plant, and animal cells, making it an ideal fluorescent marker protein. To identify the intracellular localization of a protein, the gene for that protein is fused to the GFP coding sequence, such that a fusion protein is created. Expression of this fusion protein in cells allows visualization of the protein in living cells, and (usually) in fixed cells as well. Although care must be taken to ensure that the fusion protein behaves like the wild-type protein, the GFP fusion method is extremely powerful because it allows dynamic behaviors to be observed in living cells. We will look at cells expressing either GFP alone, or a fusion of GFP to *TUB4*, a protein that is localized to the spindle pole body.

STRAINS

1-1	TSY623	*MAT*α *ade2 his3 leu2 ura3*
1-2	TSY807	*MAT***a** *his3 leu2 lys2 ura3*
1-3	TSY800	*MAT***a**/α *ADE2/ade2 his3/his3 leu2/leu2 lys2/LYS2 ura3/ura3*
1-4	TSY481	*MAT***a**/α *ADE2/ade2 his3/his3 leu2/leu2 lys2/lys2 ura3/ura3* [pTS568]
1-5	TSY514	*MAT***a** *his3 leu2 lys2 ura3* [pTS592]

PLASMIDS

pTS568	*CEN URA3* GFP
pTS592	*CEN URA3 TUB4*-GFP

PROCEDURE

Safety Notes

Formaldehyde is toxic and is a carcinogen. It is readily absorbed through the skin and is irritating to the eyes, skin, mucus membranes, and upper respiratory tract. Wear gloves and safety glasses and always work in a chemical hood.

DAPI is a possible carcinogen. It may be harmful if it is inhaled, swallowed, or absorbed through the skin. It may also cause irritation. Wear gloves, face mask, and safety glasses, and do not breathe in the dust.

Day 1.

You will be provided with cultures of strain 1-2 that were fixed during either log phase growth or stationary phase. Examine these using phase or differential interference contrast (DIC) microscopy and count 100 cells of each culture, noting the numbers of unbudded, small budded, and large budded cells.

You will also be provided with a fixed mating mixture of 1-1 and 1-2 cells. Examine this culture using phase or DIC microscopy and identify shmoos and zygotes, based on their distinctive morphology.

Compare the morphologies of haploid 1-2 and diploid 1-3 cells, looking for differences in the size and shape of cells.

Start overnight cultures of:

strain 1-3 in 10 ml of YPD at 30°C

strain 1-3 in 5 ml of YPEG (same as YPD except 3% ethanol and 3% glycerol replace glucose as the carbon source) at 30°C

strains 1-4 and 1-5 in 5 ml of SGal-ura (minimal medium with 2% galactose as carbon source) at 30°C for use the following morning.

Day 2.

Dilute back the cultures from yesterday if they are overgrown, and grow to early log phase.

(i) Fix 5 ml of the strain 1-3 cells grown in YPD by adding 0.5 ml of formaldehyde directly to culture (total concentration is 3.7%; standard stock solution is 37%). Incubate for 1–2 hours at room temperature (~23°C). Pellet cells in a clinical centrifuge and wash twice with 0.1 M potassium phosphate (pH 7.5). Store cells overnight at 4°C.

(ii) Stain living cells of strain 1-3 grown in YPD and YPEG with $DiOC_6$ according to Techniques and Protocols #10, Yeast Vital Stains.

(iii) Stain living cells of strain 1-3 grown in YPD with Calcofluor according to Techniques and Protocols #10, Yeast Vital Stains.

(iv) Make wet mounts of the strains 1-4 and 1-5 cultures as above and view the fluorescence using a fluorescein filter set.

Day 3.

(i) Stain fixed strain 1-3 YPD cells with anti-tubulin primary antibody, FITC-conjugated goat anti-rabbit secondary antibodies, and DAPI, according to Techniques and Protocols #11, Yeast Immunofluorescence.

(ii) Stain fixed 1-3 YPD cells with rhodamine phalloidin, according to Techniques and Protocols #12, Actin Staining in Fixed Cells.

MATERIALS

Day 1.
- 10 ml of YPD
- 5 ml of YPEG
- 10 ml of SGal-ura

Day 2.
- Formaldehyde (37% stock solution)
- PBS
- $DiOC_6$ (10 μg/ml stock solution in ethanol)
- Calcofluor (1 mg/ml stock solution in 100 mM potassium phosphate [pH 7.5])

Day 3.
- Zymolyase® 100T (120493-1, Seikagaku America Inc.)
- 100 mM Potassium phosphate (pH 7.5)
- Mercaptoethanol
- Polylysine
- PBS
- Methanol and acetone at –20°C
- PBS + 3% BSA
- YOL1/34 anti-tubulin antibody
- Goat anti-rat fluorescein-conjugated secondary antibody
- DAPI (Sigma D 9542 or Accurate Chemical and Scientific Corp.)
- Mounting medium
- Rhodamine-phalloidin

EXPERIMENT II

Isolation and Characterization of Auxotrophic, Temperature-sensitive, and UV-sensitive Mutants

Since spontaneous mutation frequencies are low, yeast is usually treated with such mutagens as ultraviolet (UV) radiation, nitrous acid, ethyl methanesulfonate (EMS), diethyl sulfate, and 1-methyl-nitro-nitrosoguanidine in order to enhance the frequency of mutants. These mutagens are remarkably efficient and can induce mutations at a rate of 5×10^{-4} to 1×10^{-2}/gene without substantial killing. Even though there are known methods to increase the proportion of mutants by killing off the nonmutants with nystatin and other agents, it is usually unnecessary to use selective means to obtain reasonable yields of mutants. In this experiment, auxotrophic, temperature-sensitive, and UV-sensitive mutants will be isolated from EMS-treated yeast.

Auxotrophic mutants have been invaluable for the elucidation of biochemical pathways as well as for the study of the relationship between enzyme structure and function. Studies on the intermediates accumulated by amino acid auxotrophs have facilitated the unraveling of biochemical pathways.

Studies on temperature-sensitive mutants make it possible to amplify the genetic picture of the genome. A large proportion of genes in yeast specify proteins that participate in indispensable functions (i.e., RNA polymerases, tRNA synthetases, etc.). Mutations that completely destroy the activity of these proteins are lethal and those that give only partial activity are difficult to work with genetically and biochemically. A mutation that affects the structure of one of these indispensable proteins such that it can function at low temperature but not at high temperature is much more useful. A strain carrying such a mutation grows at normal or nearly normal rates at low temperature but does not grow on any medium at elevated temperatures. This phenotype distinguishes the temperature-sensitive mutation in a vital function from a supplementable temperature-sensitive mutant (i.e., one whose defect is in the biosynthesis of an amino acid). A complete study of temperature-sensitive mutations should allow a description and understanding of the functioning of almost the entire genome. In several bacteriophages, this expectation has been realized.

The processes of DNA repair and recombination have become amenable to study through the use of UV-sensitive (or Rad^{-}) mutants. The logic behind this is that there are enzyme systems that repair UV-induced lesions of DNA. Mutation in the *RAD* genes specifying these enzymes renders the cells incapable of repairing UV-induced damage and therefore much more sensitive to radiation than wild-type cells. Because some of the UV-sensitive mutants also have impaired recombination ability, it is believed that DNA repair and recombination may take place by similar mechanisms. Also, some *RAD* genes control vital functions and their disruption is lethal.

In certain cases, it is valuable to be able to select directly for mutations in particular genes. For example, sometimes it is convenient to introduce auxotrophic markers into strains without having to put them through a genetic cross. Two particularly easy direct selections for auxotrophic mutations are the use of α-aminoadipate (αAA) for selection of *lys2* and *lys5* mutations (Chattoo et al. 1979; Zaret and Sherman 1985) and the use of 5-fluoro-orotic acid (5-FOA) for selection of *ura3* and *ura5* mutations (Boeke et al. 1986).

The standard treatment described here involves treatment of wild-type yeast strains with EMS. Half of the class will mutagenize a *MAT***a** strain, half a *MAT*α strain. After mutagenesis, the strains will be diluted and plated onto a complete medium plate at a concentration of ~100–200 cells/plate. After these cells have grown into colonies, they will be transferred to various media by replica-plating. Temperature-sensitive mutants will be detected by comparing pairs of plates that were incubated at room temperature (~23°C) and 37°C. UV-sensitive mutants will be detected by comparing irradiated plates with the controls. Auxotrophic mutants will be detected by lack of growth on a minimal medium that contains glucose, potassium phosphate, ammonium sulfate, a few vitamins, salts, and trace metals. The specific requirements can be determined by testing the colony from the original YPD plate on various types of synthetic media.

To identify auxotrophic mutants, putative mutants will be placed on a YPD plate according to a pattern. After they grow up, they will then be transferred to minimal (SD) medium plates containing pools of various amino acids, purines and pyrimidines, and other metabolites. From the pattern of growth of a particular strain, it will be possible to identify the specific requirement of the mutant strain.

Pools	#1	#2	#3	#4	#5
#6	adenine	guanine	cysteine	methionine	uracil
#7	histidine	leucine	isoleucine	valine	lysine
#8	phenylalanine	tyrosine	tryptophan	threonine	proline
#9	glutamate	serine	alanine	aspartate	arginine

Generally, a colony will respond on a plate containing one of the pools from 1–5 and on another plate containing one of the pools from 6–9, thus allowing direct identification of a single growth factor requirement. For example, a colony growing on pools 1 and 7 requires histidine or on pools 3 and 8 requires tryptophan. If a colony grows only on one of the nine pools, it requires more than one of the nutrients in that pool. A mutant blocked early in the pathway of aromatic amino acid biosynthesis will grow only on pool 8.

STRAINS

2-1	S288C	*MATα mal gal2*
2-2	D665-1A	*MAT***a**

PROCEDURE

Safety Notes

EMS is a potent mutagen. Wear gloves and work in a hood when tubes are open. Neutralize all EMS waste with 5% sodium thiosulfate before discarding. Use disposable pipettes for all manipulations.

Exposure of the skin or the eyes to UV irradiation is dangerous and should be avoided. To prevent exposure, irradiate plates in an enclosed box equipped with a trap door.

Day 1.

Inoculate 5 ml of YPD with one of the strains above (half of the class will use the *MATα* strain 2-1, while the remainder of the class will use the *MAT***a** strain 2-2). Remember to include a culture tube with medium that has not been inoculated as a control. Grow overnight at 30°C.

Day 2.

The cultures should contain ~2×10^8 cells/ml. Transfer two separate 1-ml samples of the same overnight culture to sterile microfuge tubes. Pellet the cells in a microfuge and wash them twice with sterile distilled H_2O. (Yeast cells will pellet in 5 seconds.) Resuspend the cells in 1 ml of sterile sodium phosphate buffer (pH 7). Determine the exact cell density with a counting chamber. Add 30 μl of EMS to one of the two tubes and disperse by agitation. (The other tube will be the unmutagenized control.) Incubate both tubes for 1 hour at 30°C, with agitation. Centrifuge in a microfuge and resuspend the cells in sterile H_2O. Transfer to fresh tubes, centrifuge, and wash twice with 5% sodium thiosulfate. Resuspend cells in 1 ml of sterile H_2O. The EMS treatment will cause ~40–70% inactivation. Three steps will be taken with the mutagenized culture:

1. The cells will be spread directly on αAA medium and 5-FOA medium to select for *lys2* and *ura3* mutants.
2. The cells will be grown out in YPD overnight and then spread on αAA and 5-FOA media.
3. The cells will be diluted and spread on YPD plates for colonies to be screened for other mutations.

First, spread 0.1 ml of the mutagenized cells directly on an αAA plate and on a 5-FOA plate. In addition, as a control, plate 0.1 ml of the washed, unmutagenized cells onto duplicate αAA and 5-FOA plates. The frequencies of colonies arising from the mutagenized and unmutagenized cultures will also serve to monitor the frequency of mutagenesis caused by the EMS treatment.

Second, inoculate 0.1 ml of each cell suspension (mutagenized and unmutagenized) into 1 ml of YPD broth. Grow overnight at 30°C.

Third, to obtain 100–200 viable cells/plate, dilute the EMS-treated cells (which are in the sodium thiosulfate solution) in sterile H_2O by a factor of 10^{-5}. The dilution factor may need to be adjusted, depending on the initial concentration of cells used. Spread 0.1, 0.2, and 0.4 ml each on separate YPD plates, using ten plates for each of the three different volumes plated. Incubate all of the plates for 3–4 days at room temperature (23°C).

Day 3.

Wash the 1-ml YPD overnight cultures from Day 2 with H_2O and resuspend in 1 ml of H_2O. Spread 0.1 ml onto duplicate αAA and 5-FOA plates. Incubate at 30°C.

Day 5.

Choose ten or more YPD plates containing 25–200 colonies/plate for the isolation of (1) auxotrophic mutants, (2) temperature-sensitive mutants, and (3) UV-sensitive mutants. Transfer the colonies from each of the YPD plates by replica-plating onto one SD plate, one SC plate, and four YPD plates. Record the number of colonies transferred. Make certain that each plate is numbered and has an orientation symbol on the back. For detection of auxotrophic mutants, incubate the SC and SD plates for 1 day at 30°C. For temperature-sensitive mutants, incubate for 1 day one of the YPD replicas per set at 37°C and one at room temperature (23°C). For UV-sensitive mutants, irradiate one of the YPD replicas per set with UV light (400 μJ setting in a Stratalinker) and incubate the treated and untreated plates for 1 day at 30°C. Estimate the survival after EMS.

Day 6.

Auxotrophic Mutants. Compare each of the ten SD plates with the SC plates. Using sterile toothpicks, transfer by frogging colonies whose replicates have failed to grow on the SD plates onto the nine plates with the various pools of metabolites, and onto an SC and an SD plate. There should be 5% of such colonies. Include also strain 2-1 or 2-2. Incubate overnight at 30°C.

UV-sensitive Mutants. Compare the irradiated plates with the unirradiated controls. Transfer by frogging the colonies not appearing on the irradiated plate onto two YPD plates. Include strain 2-1 or 2-2 also. Irradiate one of the plates as before and incubate the two plates overnight at 30°C.

Temperature-sensitive Mutants. Compare the 37°C plate with the 23°C plate and transfer by frogging any colonies that failed to grow at the high temperature onto two YPD plates. Incubate for 1 day one plate at 37°C and the other at room temperature. (Also include strain 2-1 or 2-2.) A replica plate may also be made onto YPD containing 30% sucrose. This plate will allow those temperature-sensitive mutants to grow at 37°C whose defects cause osmotic lysis on the YPD plate or whose defects are suppressed by high osmotic conditions (osmotic-remedial mutants).

lys2 *and* ura3 *Mutants.* Count the number of colonies arising on the αAA and 5-FOA plates spread on Days 2 and 3. Estimate the frequency of each type of mutant before and after mutagenesis and before and after outgrowth following mutagenesis, taking 2×10^8 cells/ml as the saturation density of your YPD overnight cultures. Purify four colonies from an αAA plate and four from a 5-FOA plate on YPD to check their Lys and Ura phenotypes. Incubate the plates at 30°C. Was the frequency of each type of mutant increased after mutagenesis? Why might growth after mutagenesis affect the recovery of each type of mutant?

Days 7–8.

Auxotrophic Mutants. Record the growth response and assign a number to each mutant. Make a streak master of your mutants on a YPD plate. Incubate overnight at 30°C.

UV-sensitive Mutants. Record the growth response and assign a number to each mutant. These strains may also be checked for other aspects of their phenotypes—sensitivity to X-rays and alkylating agents as well as rates of reversion and recombination of selected markers. Make a streak master of your mutants on a YPD plate. Incubate at 30°C.

Temperature-sensitive Mutants. Record the degrees of growth of your mutants at 23°C and 37°C. Assign a number to each mutant. Make a streak master of your mutants on a YPD plate. Incubate overnight at room temperature.

lys2 *and* ura3 *Mutants.* Frog the purified αAA^R colonies onto SD and SD + lys plates to test for the Lys phenotypes. Frog the purified 5-FOA^R colonies onto SD and SD + ura plates to test the Ura phenotype.

Day 9.

Exchange mutants of similar phenotypes with people who isolated their mutants from strains having the opposite mating type. Cross the mutants on YPD plates. Incubate overnight at room temperature.

Day 10.

Auxotrophic Mutants. Replica-plate the cross-stamped plates onto SD plates for complementation testing. Incubate for 1–2 days at room temperature.

UV-sensitive Mutants. Replica-plate the cross-stamped plate onto a YPD plate, and irradiate. A diploid will have normal UV sensitivity if it was formed from two recessive Rad^- strains that had mutations at different loci. Incubate for 1–2 days at room temperature.

Temperature-sensitive Mutants. Replica-plate the cross-stamped plate onto a YPD plate and incubate for 1–2 days at 37°C.

Day 12.

Score the complementation tests. Record the data from the entire class and determine the minimum number of genes that control each character.

lys2 *and* ura3 *Mutants.* Record the growth of the *lys2* and *ura3* mutants from Day 7–8.

MATERIALS

***Note:* Amounts provided are the requirements for each pair.**

Day 1.	2 Culture tubes, each containing 5 ml of YPD
Day 2.	Sterile distilled H_2O 5 ml of Sterile 0.1 M sodium phosphate buffer (pH 7) EMS (Methanesulfonic acid ethyl ester; Sigma M 0880) Sterile 5% sodium thiosulfate (w/v) 2 Culture tubes, each containing 1 ml of YPD 2 αAA plates 2 5-FOA plates 30 YPD plates
Day 3.	2 αAA plates 2 5-FOA plates
Day 5.	10 SC plates 10 SD plates 40 YPD plates Sterile velveteen pads
Day 6.	~11 YPD plates ~2 each of the nine pool plates ~2 SD plates ~4 YPD plates 2 YPD plates containing 30% sucrose Sterile velveteen pads 2 SC plates
Day 7–8.	~6 YPD plates 2 SD plates 1 SD + lys plate 1 SD + ura plate
Days 9–11.	~5 YPD plates ~5 SD plates Sterile velveteen pads

REFERENCES

Auxotrophic Mutants

Lindegren, G., L.Y. Hwang, Y. Oshima, and C. Lindegren. 1965. Genetical mutants induced by ethyl methanesulfonate in *Saccharomyces. Can. J. Genet. Cytol.* **7:** 491–499.

Lingens, F. and O. Oltmanns. 1964. Erzeugung und untersuchung Biochemischer and Mangelmutanten von *Saccharomyces cerevisiae. Z. Naturforsch.* **19B:** 1058–1065.

———. 1966. Uber die Mutagene Wirkung von 1-nitroso-3-nitro-1-methyl-guanidin (NNMG) und *Saccharomyces cerevisiae. Z. Naturforsch.* **21B:** 660–663.

Temperature-sensitive Mutants

Hartwell, L.H. 1967. Macromolecule synthesis in temperature-sensitive mutants of yeast. *J. Bacteriol.* **93:** 1662–1670.

Pringle, J.R. and L.H. Hartwell. 1981. The *Saccharomyces cerevisiae* cell cycle. In *The molecular biology of the yeast* Saccharomyces: *Life cycle and inheritance* (ed. J.N. Strathern et al.), pp. 97–142. Cold Spring Harbor Laboratory, Cold Spring Harbor, New York.

UV-sensitive Mutants

Cox, B.S. and J.M. Parry. 1968. The isolation, genetics and survival characteristics of ultraviolet-light-sensitive mutants in yeast. *Mutat. Res.* **6:** 37–55.

Game, J.C. and B.S. Cox. 1971. Allelism tests of mutants affecting sensitivity to radiation in yeast and a proposed nomenclature. *Mutat. Res.* **12:** 328–331.

Moustacchi, E. 1972. Evidence for nucleus independent steps in control of repair of mitochondrial damage. I. UV-induction of the cytoplasmic "petite" mutation in UV-sensitive nuclear mutants of *Saccharomyces cerevisiae. Mol. Gen. Genet.* **114:** 50–58.

Resnick, M.A. 1969. Genetic control of radiation sensitivity in *Saccharomyces cerevisiae. Genetics* **62:** 519–531.

Selections for Auxotrophs

Boeke, J.D., F. LaCroute, and G.R. Fink. 1986. A positive selection for mutants lacking orotidine-5′-phosphate decarboxylase activity in yeast; 5′-fluoro-orotic acid resistance. *Mol. Gen. Genet.* **197:** 345–346.

Chattoo, B.B., F. Sherman, D.A. Azubalis, T.A. Fjellstedt, D. Mehnert, and M. Ogur. 1979. Selection of *lys2* mutants of the yeast *Saccharomyces cerevisiae* by the utilization of α-aminoadipate. *Genetics* **93:** 51–65.

Zaret, K.S. and F. Sherman. 1985. α-Aminoadipate as a primary nitrogen source for *Saccharomyces cerevisiae* mutants of yeast. *J. Bacteriol.* **162:** 579–583.

Enrichment Methods

Henry, S.A., T.F. Donahue, and M.R. Culbertson. 1975. Selection of spontaneous mutants by inositol starvation in *Saccharomyces cerevisiae. Mol. Gen. Genet.* **143:** 5–11.

Snow, R. 1966. An enrichment method for auxotrophic yeast mutants using the antibiotic "nystatin." *Nature* **211:** 206–207.

Thouvenot, D.R. and C.M. Bourgeois. 1971. Optimisation de la selection de mutants de *Saccharomyces cerevisiae* par la nystatine. *Ann. Inst. Pasteur* **120:** 617–625.

Walton, B.F., B.L.A. Carter, and J.R. Pringle. 1979. An enrichment method for temperature-sensitive and auxotrophic mutants of yeast. *Mol. Gen. Genet.* **171:** 111–114.

Experiment III

Meiotic Mapping

Linkage studies are possible by both meiotic mapping (tetrad analysis) and mitotic mapping. The four spores in an ascus are the products of a single meiotic event, and the genetic analysis of these tetrads can provide linkage relationships of genes present in the heterozygous condition. It is also possible to map a gene relative to its centromere if known centromere-linked genes are present in the hybrid. Although the isolation of four spores from an ascus is one of the more difficult techniques, requiring a micromanipulator and considerable practice, tetrad analysis is useful not only for linkage studies but also for constructing strains necessary in genetic and biochemical experiments. In some cases, spores can be isolated randomly (see Experiment IV for an illustration of the technique).

The three classes of tetrads—parental ditype (PD), nonparental ditype (NPD), and tetratype (T)—from a diploid that is heterozygous for two markers, *AB* x *ab*, are given in the following table:

PD	NPD	T
AB	*aB*	*AB*
AB	*aB*	*Ab*
ab	*Ab*	*ab*
ab	*Ab*	*aB*

	PD		NPD		T
Random assortment	1	:	1	:	4
Linkage	>1	:	<1		
Centromere linkage	1	:	1	:	<4

There is an excess of PD to NPD asci if two genes are linked. If two genes are on different chromosomes and are linked to their respective centromeres, there is a reduction of the proportion of T asci. If two genes are on different chromosomes and at least one gene is not centromere-linked or if two genes are widely separated on the same chromosome, there is independent assortment and the PD : NPD : T ratio is 1 : 1 : 4.

Since the NPD tetrads have four recombinants and the T tetrads have only two recombinants, the percentage of recombination is equal to:

$$100 \frac{1/2(\mathrm{T}) + \mathrm{NPD}}{\text{Total tetrads}} = 100 \frac{\mathrm{T} + 2(\mathrm{NPD})}{2(\mathrm{PD} + \mathrm{NPD} + \mathrm{T})}$$

This value is equal to the distance in centiMorgans between the two genes if there is a maximum of one exchange within the interval (i.e., if there are no NPD asci). (The distance is equal to half the percentage of T asci.) If the proportion of NPD asci is small and there are two or less exchanges between the linked markers, the distance in centiMorgans is approximately:

$$100 \frac{1/2\,(\mathrm{T} - 2[\mathrm{NPD}]) + 4(\mathrm{NPD})}{\text{Total tetrads}} = 100 \frac{\mathrm{T} + 6(\mathrm{NPD})}{2(\mathrm{PD} + \mathrm{NPD} + \mathrm{T})}$$

There are four types of double-crossover progeny. One type yields PD, T, and the fourth yields NPD. Since the only unique class is the NPD, the frequency of all double crossovers, assuming no chromatid interference, can be inferred to be 4(NPD). A correction has to be made for the single crossovers by subtracting the double crossovers that yield tetrads—hence the term T – 2(NPD).

A number of assumptions concerning interference must be made to determine map distances between larger intervals. In addition, the calculations become inaccurate as the distances approach the limiting values, and the only accurate way of measuring long intervals is by the summation of shorter intervals.

The distance from a gene to its centromere can be measured by the frequency of second-division segregation (SDS), which is analogous to the frequency of T asci of two gene markers. The distance in centiMorgans is equal to half of the percentage of SDS if the interval between the gene and its centromere is small. The percentage of SDS for a particular gene can be easily determined if the hybrid contains a closely linked centromere marker such as *trp1*, which is approximately one centiMorgan (map unit) from its centromere. In this case, the percentage of SDS is equal to the percentage of T asci with an error of less than 1%. This is because *trp1* segregates at the first meiotic division with its centromere. It is also possible to determine the percentage of SDS with reasonable accuracy if the hybrid contains two or more centromere-linked markers that may not be as close as, for example, *trp1*. In this case, the SDS array is decided upon by the best agreement among the centromere-linked markers. It should also be noted that the SDS frequencies of two loci, A and B, on different chromosomes are related to the frequency of tetratype asci $(\mathrm{T})_{\mathrm{AB}}$ by:

$$(\mathrm{T})_{\mathrm{AB}} = (\mathrm{SDS})_{\mathrm{A}} + (\mathrm{SDS})_{\mathrm{B}} - 3/2\,(\mathrm{SDS})_{\mathrm{A}}\,(\mathrm{SDS})_{\mathrm{B}}$$

However, this relationship is only reasonably accurate if at least one of the markers is fairly near its respective centromere.

Meiosis and sporulation usually occur within 4 days when diploid cells are transferred to sporulation medium. A sporulated culture contains a mixture of unsporulated diploid cells, of asci with four haploid spores, and of asci with less than four spores. The spores will not germinate or divide on sporulation medium, and most sporulated cultures may be stored in the refrigerator for several months without loss of viability. The first step in dissection consists of treating the sporulated culture with Zymolyase, which digests the ascus wall but does not disturb the association of the four spores from the ascus. Digested asci are carefully transferred onto a YPD plate using a sterile incubating loop. The four spores are then separated from each other and moved to isolated positions on the YPD plate using the micromanipulator. The spores will germinate and form colonies in 2–3 days after dissection. The spore clones can be transferred to slants or to YPD plates for storage. A master plate is prepared by inoculating a YPD plate with the strains; most of the phenotypes are scored by replica-plating onto appropriate media. The mating types, complementary markers, and particular alleles within a single gene can be determined by replica-plating the master plate onto two YPD plates and replica-plating onto these prints with lawns of *MAT***a** and *MATα* tester strains. The YPD plates containing the mating mixtures are replica-plated onto minimal media after 1 day of incubation.

Many of the techniques and principles of chromosome mapping will be illustrated by tetrad analysis of the diploid 3-3, which was constructed by a cross of strains 3-1 and 3-2. A sporulated culture of the above strain will be furnished. Since micromanipulation requires considerable practice, the schedule will commence after a number of asci have been successfully dissected.

STRAINS

3-1	1385-3B	*MAT***a** *his4-G met13 lys2 ho::LYS2*
3-2	1384-10D	*MATα his4-N met4 ura3 leu2 trp1 lys2 ho::LYS2*
3-3	1385-3B x 1384-10D	
3-4	FW786	*MAT***a** *ade8*
3-5	FW787	*MATα ade8*
3-10	ERX-17C	*MAT***a** *his4-N met4 ade8*

3-11	ERX-3C	*MAT*α *his4-N met4 ade8*
3-12	ER3**a**	*MAT***a** *his4-G met13 ade8*
3-13	ER3α	*MAT*α *his4-G met13 ade8*

Note: Strain 3-3 will sporulate after storage on YPD. The diploid may be reconstructed by mating strains 3-1 and 3-2.

PROCEDURE

Day 1.

Preparation of the Microneedle. The separation of ascospores is carried out with glass microneedles attached to any one of a variety of micromanipulators (Sherman 1975). A simple microneedle for separating ascospores (as well as cells and zygotes) can be easily made using the small flame from the pilot light of an ordinary Bunsen burner. An even easier method for making dissecting needles has recently become popular; it requires some Super Glue and some fiber optic glass. Three methods are described below. The third (and by far the easiest) method will be used in the course.

METHOD 1. A 2-mm-diameter glass rod is drawn out to a fine tip using the Bunsen burner; by using the pilot burner, an even finer tip is drawn out at a right angle with an auxillary piece of glass rod as illustrated in Figure 1. The end is broken off so that the tip has a diameter of 10–100 μm and a length of a few millimeters. The exact diameter is not critical and various investigators have different preferences. Spores are more readily picked up and transferred with microneedles having tips of larger diameters, whereas manipulations in crowded areas having high densities of cells are more manageable with microneedles having smaller diameters. A needle with a diameter of ~40 μm is an acceptable compromise. The length of the perpendicular end should be compatible with the height of the chamber—too short an end may result in optical distortions from the main shank of the microneedle; longer microneedles are required for manipulations on the surface of petri dishes. The drawn-out tip can be cut with a razor blade or broken between the surface and edge of two glass slides. It is critical that the microneedle should have a flat end, which sometimes requires several attempts.

METHOD 2. Microneedles can also be made by drawing thin filaments from a 2-mm-diameter glass rod using a small gas flame (Scott and Snow 1978). Segments that appear to be the correct diameter are chopped into lengths of ~2 cm with a razor blade as illustrated in Figure 2. The short segments are inspected with a microscope to determine if they have the correct diameter and a flat end. The end of a 2-mm-diameter glass rod is then tapered and bent to a right angle. A drop of Super Glue is applied to this end of the mounting rod, which is then touched to one of the segments of a glass

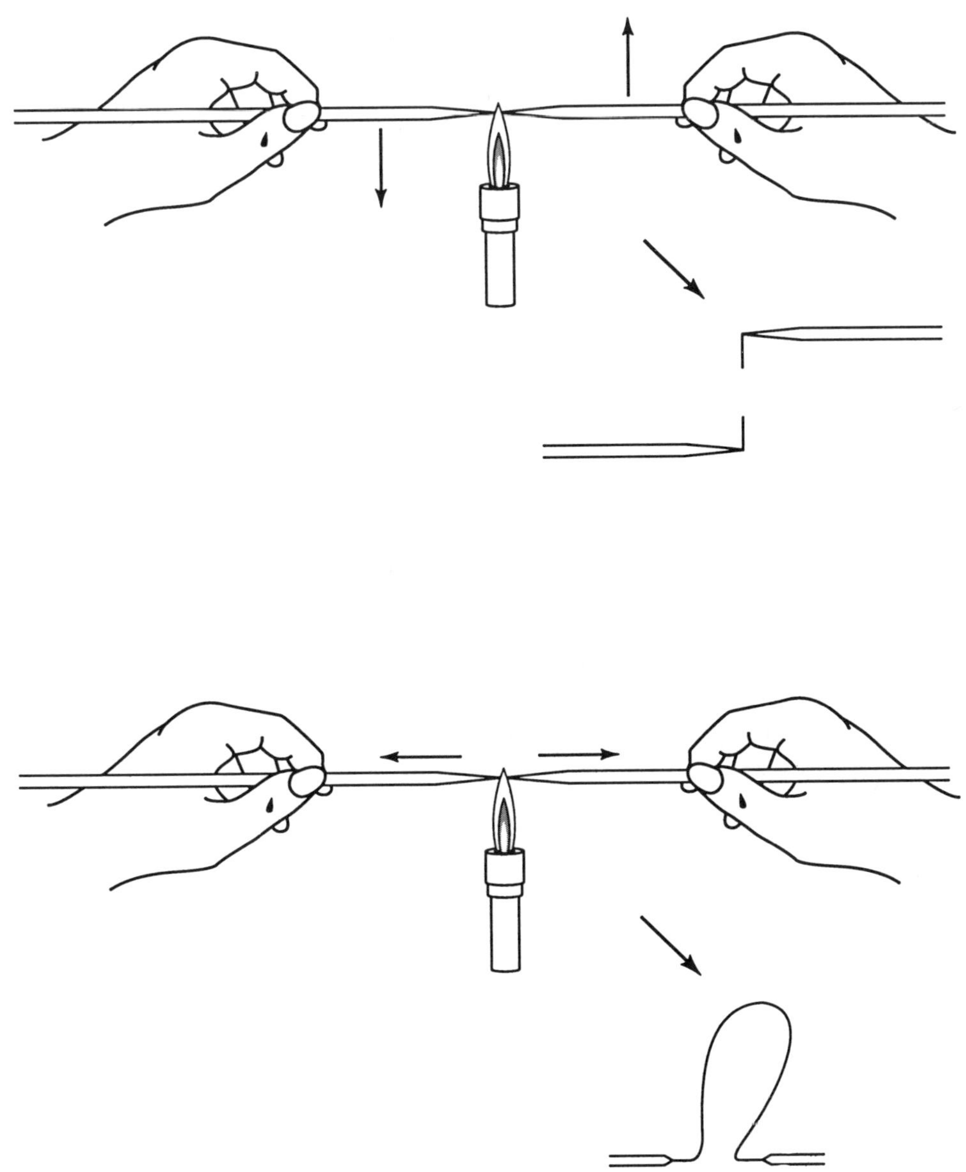

Figure 1 Construction of microneedles. (*Top*) Microneedles required for the separation of ascospores can be made by first drawing out a 2-mm-diameter glass rod into a fine tip and then drawing out the end to an even finer tip at a right angle. (*Bottom*) The right-angle end of the microneedle also can be made by looping-out the fine section.

filament as shown in Figure 2. The filament is carefully positioned so that it is at a right angle to the axis of the mounting rod and so that it has the correct length. The microneedle is fitted into the micromanipulator after the mountant has dried. This procedure is especially useful if numerous needles are required.

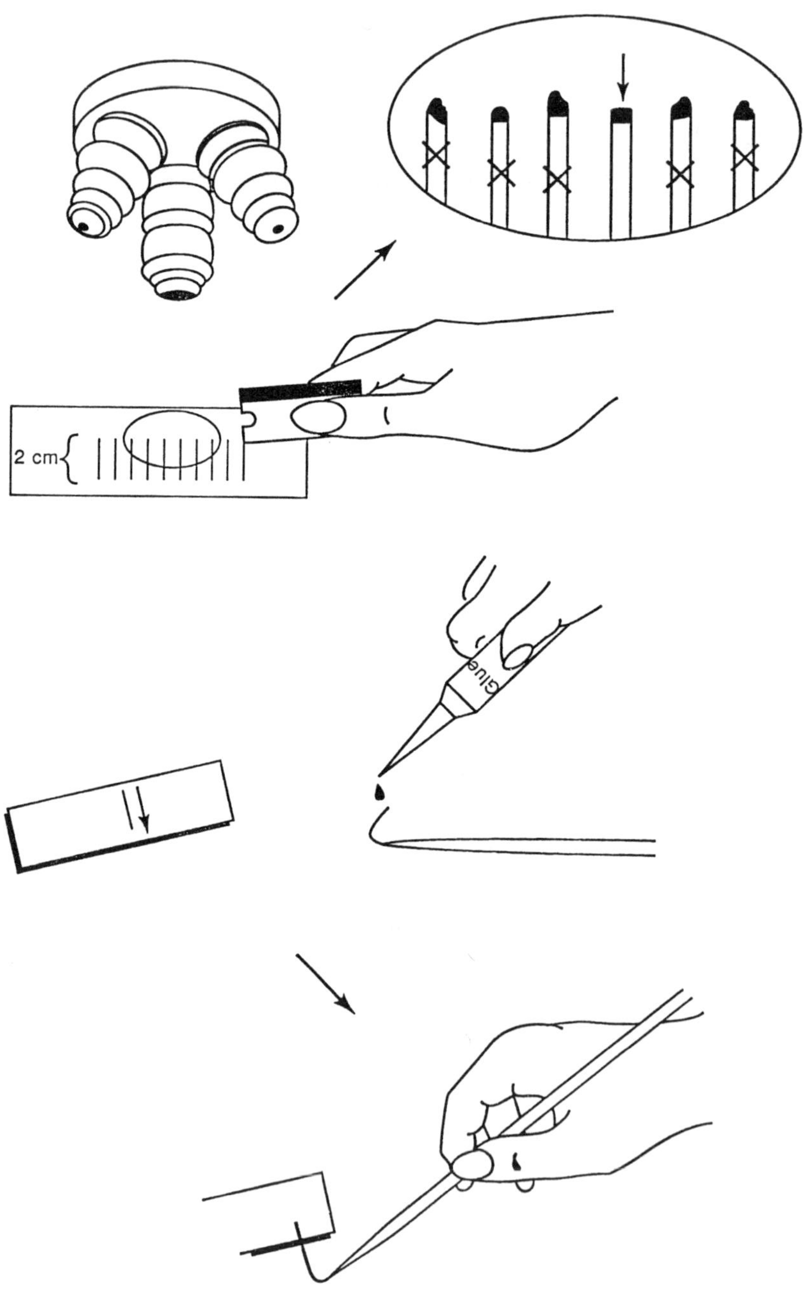

Figure 2 Use of glass filaments. Microneedles also can be made with thin glass filaments ~40 μm in diameter. Segments ~2 cm long are broken with a razor blade and examined with a microscope. The segments containing flat ends are attached at a right angle to a mounting rod with Super Glue.

METHOD 3. This method is the same as Method 2 except that instead of pulling out a needle, commercially available fiber optic glass of the correct diameter is used. The glass fiber can be broken into several pieces, inspected for those pieces with a flat end, and Super Glued onto a glass rod base as in Method 2. A flat end is most easily obtained by breaking the end of the glass fiber with a cover slip by holding the cover slip at a 45° angle (like a tiddleywink) to chop the end of the fiber. The instructors will demonstrate this technique. You should then mount the microneedle on a micromanipulator and move the microneedle to various positions until you become familiar with the controls.

Treatment with Zymolyase. Examine the sporulated culture under a microscope and identify the unsporulated cells, the four-spored asci, and so on. Using the flat end of a sterile toothpick, resuspend a small dab of the sporulated culture in 50 μl of Zymolyase solution (0.5 mg/ml in 1 M sorbitol) and incubate for 10 minutes at 30°C. Slowly and carefully add 0.8 ml of sterile distilled H_2O to the tube and place it on ice. The asci should now be ready for dissection. Transfer a drop of the asci onto a YPD plate and spread them in a single line near the edge of the plate.

Dissection of Asci. Place the YPD plate containing the digested asci on the microcope stage; be extremely careful not to break the microneedle. Move the chamber with the rack and pinion so that the microneedle is ~2 mm from the agar and over the streak of spores. Examine the spores in the vicinity and select a cluster of four spores. Pick up the four spores with the microneedle and place them on the agar ~5 mm from the line of asci. Note the position on the mechanical stage. Pick up three spores and move the chamber 5 mm away from the streak. Plant the three spores and pick up two spores. Move the chamber 5 mm away from the streak, plant the two spores, and pick up one spore. Move the chamber 5 mm and plant the remaining spore. Move the chamber 5 mm from the line of the four spores and select another four-spore cluster. Separate the spores as before by 5-mm intervals. By this method, ten tetrads can be dissected on each side of the YPD plate. Remove the YPD plate from the stage, taking care not to break the microneedle. Incubate the plate for 2–3 days at 30°C.

Day 2.

Preparation of Mating-type, Complementation, and Allele Testers. Inoculate 5 ml of YPD with the mating-type testers 3-4 and 3-5 and with the following *his4*, *met4*, and *met13* tester strains:

3-10	ERX-17C	*MAT***a** *his4-N met4 ade8*
3-11	ERX-3C	*MAT*α *his4-N met4 ade8*
3-12	ER3**a**	*MAT***a** *his4-G met13 ade8*
3-13	ER3α	*MAT*α *his4-G met13 ade8*

Grow for 1 day at 30°C.

Day 3.

Using sterile toothpicks and following the grid provided (see Appendix D), prepare a master plate with spore colonies that have four viable members. Include the two parental control strains on each plate. Ten complete tetrads plus the controls can be regridded on one plate. Incubate for 1 day at 30°C.

For each mating-type, complementation, and allele tester, spread 0.2 ml of the overnight culture on a YPD plate and incubate for 1 day at 30°C.

Day 4.

Replica-plate the master plate to these media: SC – his, SC – met – cys, SC – ura, SC – leu, SC – trp, SC, and seven YPD plates. One YPD is a positive control for growth. Replica-plate one tester lawn to each of the six other YPD plates. Incubate all plates at 30°C.

(Alternatively, sterile H_2O in a microtiter dish can be inoculated with spore colonies and the cells can be transferred to the different test plates using the multiprong inoculator [frogging technique]. This will be demonstrated by the instructors.)

Day 5.

Replica-plate the matings to strains 3-4 and 3-5 to SD plates. Replica-plate the other matings to SD + his + met and to SD + his. Incubate all plates at 30°C.

Day 7.

Score growth on the SD + his + met and SD + his plates. Replica-plate the SD + his + met plates to an SC – his plate and to an Spo plate. Incubate the replicas at 30°C. Score auxotrophic markers from the plates of Day 4, and mating types from the plates of Day 5.

Day 9.

Replica-plate the Spo plates to SC – his, and incubate at 30°C. Refrigerate your SC – his replicas from Day 7.

Day 11.

Compare growth on the SC – his replicas of Day 7 and Day 9. Score presence of His$^+$ recombinants on your replicas from Day 9.

MATERIALS

***Note:* Amounts provided are for 20 tetrads—double the quantities given per group.**

Day 1.

2-mm-diameter Glass rod for preparing microneedles
Super Glue®
Fiber optic glass
Zymolyase® 100T (120493-1, Seikagaku America Inc.) solution (0.5 mg/ml in 1 M sorbitol)
Sterile distilled H_2O
1 YPD plate

Day 2.

6 Culture tubes, each containing 5 ml of YPD

Day 3.

8 YPD plates

Day 4.

2 Plates each of:
- SC – his
- SC – met – cys
- SC – ura
- SC – leu
- SC – trp
- SC

14 YPD plates
Sterile velveteen pads

Day 5.

4 SD plates
8 SD + his plates
8 SD + his + met plates
Sterile velveteen pads

Day 7.

8 SC – his plates
8 Spo plates
Sterile velveteen pads

Day 9. 8 SC – his plates
Sterile velveteen pads

REFERENCES

The Fincham et al. (1979) reference cited in the Introduction and Mortimer and Hawthorne (1969) give general information concerning meiotic mapping and tetrad analysis.

Mortimer, R.K. and D.C. Hawthorne. 1966. Genetic mapping in *Saccharomyces. Genetics* **53:** 165–173.

———. 1969. Yeast genetics. In *The yeasts* (ed. A.H. Rose and J.S. Harrison), vol. 1, pp. 385–460. Academic Press, New York.

Mortimer, R.K. and D. Schild. 1981. Genetic mapping in *Saccharomyces cerevisiae.* In *The molecular biology of the yeast* Saccharomyces: *Life cycle and inheritance* (ed. J.N. Strathern et al.), pp. 11–26. Cold Spring Harbor Laboratory, Cold Spring Harbor, New York.

Perkins, D.D. 1949. Biochemical mutants in the sput fungus *Ustilago maydis. Genetics* **34:** 607–626.

Scott, K.E. and R. Snow. 1978. A rapid method for making glass micromanipulator needles for use with microbial cells. *J. Gen. Appl. Microbiol.* **24:** 295–296.

Sherman, F. 1975. Use of micromanipulators in yeast studies. *Methods Cell Biol.* **11:** 189–199.

Experiment III: Tetrad analysis; 3-3 Group no.______

Names____

		MAT	*his4*	*leu2*	*ura3*	*trp1*	*met4*	*met13*
MAT.....	P							
	N							
	T							
his4.....	P							
	N							
	T							
leu2....	P							
	N							
	T							
ura3.....	P							
	N							
	T							
trp1.....	P							
	N							
	T							
met4....	P							
	N							
	T							

% SDS.... (with *trp1*)							

EXPERIMENT IV

Mitotic Recombination and Random Spore Analysis

In diploid strains, alterations leading to loss of genetic information can be generated by mitotic crossing over, gene conversion, or chromosome loss. Because such events are usually observed as the loss of one of two alleles in a heterozygous strain, they are referred to as loss of heterozygosity (LOH) events. Mitotic crossing over and gene conversion are illustrated in Figure 1. Mitotic crossing over results in LOH of all markers located distal to the point of exchange on the chromosome arm. Mitotic gene conversion results in nonreciprocal exchange of only a small chromosomal region (i.e., a single gene or closely linked genes) and is believed to be analogous to the irregular segregations that are observed at low frequency after meiosis (meiotic gene conversion). Chromosome loss, which results in LOH of markers on both arms of a chromosome, produces a 2N-1 diploid that may or may not have impaired growth, depending on the particular chromosome.

Mitotic crossing over between two non-sister chromatids is likely to occur during the G2 stage of the cell cycle (see Fig. 1). After mitosis, a crossover event can lead to two daughter cells that are homozygous for part of the chromosome arm that was heterozygous in the mother cell (this will occur in half of the recombination events because of the random orientation of chromosomes on the mitotic spindle). Thus, mitotic recombination can lead to the ''uncovering'' of recessive markers. For some markers, ''papillae'' or ''sectors'' derived from the homozygous daughter cells can be observed in a background of heterozygous cells, resulting in papillating or sectoring colonies. As a result of mitotic crossing over and subsequent mitotic segregation, all markers distal to the site of recombination on the chromosome arm are simultaneously homozygosed. It is possible to generate a recombinational map of the chromosomal arm using mitotic recombination data. There is an approximately linear relationship between the frequency of homozygosis of a gene and its distance from the centromere. Thus, quantitation of the relative frequencies of different classes of recombinants can be used to determine the relative distances between two markers and between a marker and its centromere.

Mitotic gene conversion also results from recombination between two non-sister chromatids in G2 (see Fig. 1). In this case, only a limited chromosomal region is homozygosed, with the result that distal markers retaining their heterozygosity. If two or more markers sector together, this event most likely is due to mitotic crossing over and not to gene conversion, because conversion tracts are generally short (several kilobases at the most).

It is important to note that the spontaneous rate of mitotic recombination is low, ranging from $\sim10^{-6}$ to $\sim10^{-4}$, depending on the distance of a particular gene from its centromere and other less well-defined factors. Chromosome loss events occur at similarly low frequencies. To find rare LOH events easily, it is usually necessary to apply a selection or screen. In this experiment, we will use selections for canavanine resistance and 5-fluoro-orotic acid (5-FOA) resistance to identify cells in which a LOH event has occurred.

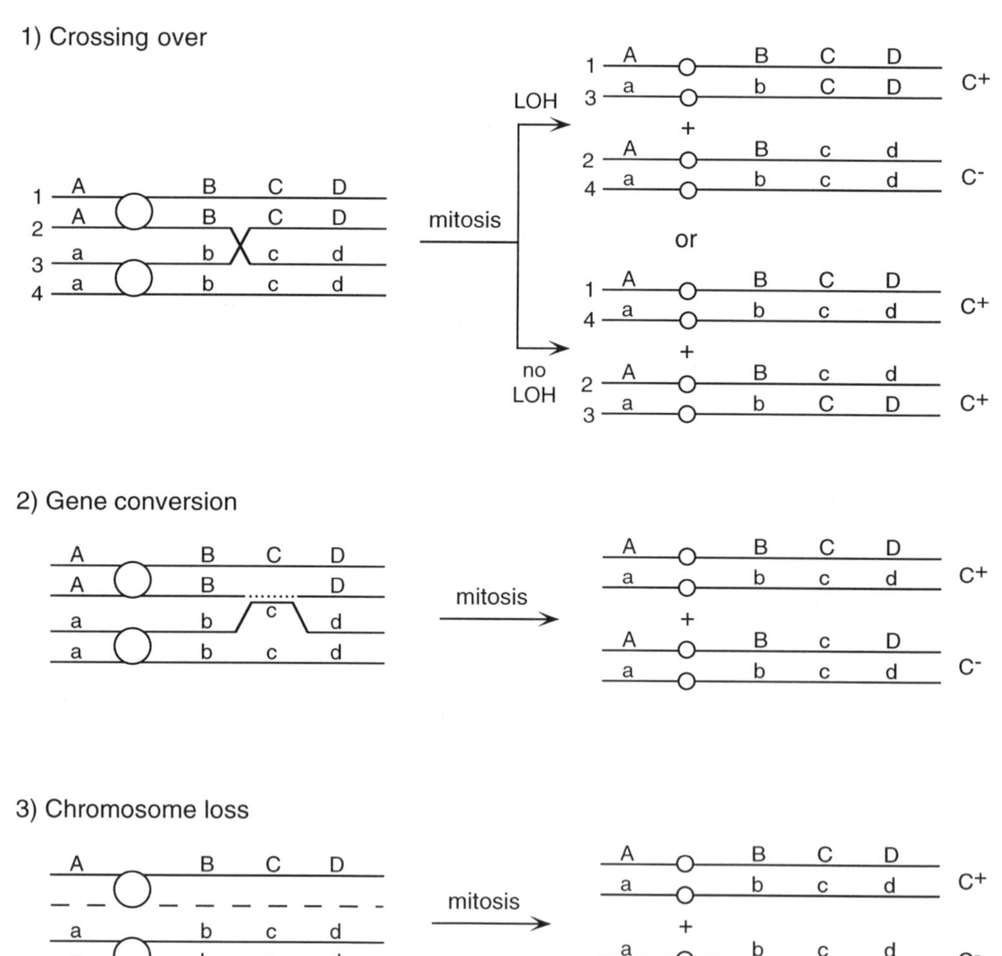

Figure 1 Mitotic segregation of heterozygous markers in a diploid strain after either (1) crossing over; (2) mitotic gene conversion; or (3) chromosome loss.

EXPERIMENTAL DESIGN

A diploid will be created by mating the following two haploid strains:

4-1 TSY812 *MATα* *can1* *hom3* *leu2* *lys2* *ura3*

4-2 TSY813 *MAT***a** *ade2* *his1* *lys2* *trp1*

URA3 and *CAN1* are on the left arm of chromosome V, and *HIS1* and *HOM3* are on the right arm of chromosome V. The *URA3* and *CAN1* genes are particularly useful for this experiment because they have the unusual feature that strong selection techniques exist for <u>recessive</u> alleles of both genes. Starting with a diploid that is heterozygous at both loci, *ura3* strains that arise by mitotic recombination, chromosome loss or meiosis can be selected by their resistance to 5-FOA, and *can1* strains by their resistance to canavanine.

The *URA3* gene encodes orotidine-5′-phosphate decarboxylase, an enzyme that catalyzes one of the steps in pyrimidine synthesis. 5-FOA is taken up by cells and is converted to the toxic compound 5-fluorouracil by the action of the decarboxylase. *URA3/URA3* homozygotes, *URA3/ura3* heterozygotes, and *URA3/* hemizygotes (one copy of gene in a diploid strain) are all able to convert 5-FOA to 5-fluorouracil, and are sensitive to 5-FOA (5-FOA^S), whereas *ura3/ura3* homozygotes and *ura3/* hemizygotes lack the decarboxylase activity and are resistant to 5-FOA (5-FOA^R).

The *CAN1* gene encodes the arginine permease, which allows uptake of arginine from the medium. Canavanine is a toxic analog of arginine that is taken up by cells through the arginine permease. *CAN1/CAN1* homozygotes, *CAN1/can1* heterozygotes, and *CAN1/* hemizygotes are all able to take up canavanine, and so are sensitive to canavanine (Can^S), whereas *can1/can1* homozygotes and *can1* hemizygotes lack the permease and are resistant to canavanine (Can^R).

The frequency of LOH at each locus will be determined by determining the frequency of Can^R and 5-FOA^R mitotic segregants that arise from the diploid strain 4-1 x 4-2. These frequencies will be compared to the frequency of spontaneous mutations in *CAN1* and *URA3* in the haploid strain 4-2. Mitotic recombination will be distinguished from chromosome loss by examining the Hom3 phenotype on the opposite arm of chromosome V. The relative positions of *CAN1* and *URA3* with respect to the centromere on chromosome V will be determined by seeing how often LOH occurs at *CAN1* but not at *URA3*, or at *URA3* but not at *CAN1*.

An additional part of this experiment will be to examine meiotic recombination in the 4-1 x 4-2 diploid using random spore analysis. This method allows meiotic products to be isolated and scored without performing tetrad analysis. It is particularly useful when many crosses must be analyzed, or a rare recombinant must be identified. It is important to realize, however, that information about centromere linkage and the ability to use the tetrad mapping function for more accurate determination of genetic distance are lost when random spore analysis is used in place of tetrad analysis. The principle behind the method is that haploid meiotic segregants are selected away from unsporulated diploid cells by selecting for canavanine resistance in a strain that was originally a *CAN1/can1* heterozygote. Half of all meiotic segregants will be Can^R because they will have received the *can1* gene.

STRAINS

4-1	TSY812	*MATα*	*can1*	*hom3*	*leu2*	*lys2*	*ura3*
4-2	TSY813	*MAT***a**	*ade2*	*his1*	*lys2*	*trp1*	

PROCEDURE

Day 1.

Subclone strains 4-1 and 4-2 to a YPD plate. Incubate at 30°C.

Day 3.

In the morning, mate strains 4-1 and 4-2. Transfer a single colony of 4-1 and a single colony of 4-2 to a YPD plate and mix together in a patch (a drop of sterile H_2O will facilitate mixing). Incubate at least 5 hours at 30°C.

In the afternoon, streak out cells from the mixed patch to an SD+lys plate to select for diploids. Incubate at 30°C.

Day 5.

(i) *LOH.* In the evening, inoculate 5 ml of YPD with the 4-1 x 4-2 diploid and 5 ml of YPD with the 4-2 haploid as a control. Incubate on the roller drum at 30°C.

(ii) *Meiotic recombination.* To prepare the 4-1 x 4-2 diploid for sporulation, patch onto a YPD plate and incubate overnight at 30°C.

Day 6.

(i) *Mutation and LOH.* Prepare tenfold serial dilutions of the saturated culture of the 4-1 x 4-2 diploid in sterile microfuge tubes. To determine the frequency of LOH, plate 100 μl of 10^{-1}, 10^{-2}, and 10^{-3} dilutions on SC+5-FOA and SC-arg+canavanine plates. To determine the number of viable cells, plate 100 μl of 10^{-4} and 10^{-5} dilutions on YPD plates. To determine the frequency of spontaneous mutation at the *CAN1* and *URA3* genes, prepare similar dilutions of the 4-2 haploid culture. For these cells, plate 100 μl of 10^{-1} and 10^{-2} dilutions on SC+5-FOA and SC-arg+canavanine plates. To determine the number of viable cells, plate 100 μl of 10^{-4} and 10^{-5} dilutions on YPD plates. Incubate all plates at 30°C.

(ii) *Meiotic recombination.* Transfer the 4-1 x 4-2 diploid from the YPD plate to liquid sporulation medium and incubate at 25°C according to Techniques and Protocols #9, Random Spore Analysis.

Day 9.

(i) *Mutation and LOH.* Count the colonies on the selective plates and YPD plates and calculate the frequencies of Can^R and 5-FOA^R for the 4-1 x 4-2 diploid and for the 4-2 haploid. With the multipronged inoculating device (frogger), spot 22 independent isolates of each type of the drug-resistant strains onto SC+5-FOA, SC-arg+canavanine, SC-met (to score *HOM3*), and YPD plates. If you don't have enough 5-FOA^R or Can^R isolates of strain 4-2, just plate what you have. Include as controls the 4-1 x 4-2 diploid and the parental haploids 4-1 and 4-2.

(ii) *Meiotic recombination.* Check the sporulating culture of 4-1 x 4-2. If the culture has sporulated, digest and break apart tetrads, and plate dilutions of the random spores according to Techniques and Protocols #9, Random Spore Analysis.

Day 10.

Mutation and LOH. Based on growth of patched or spotted strains, calculate the frequency of Can^R for strains first selected on 5-FOA and 5-FOA^R for strains first selected on canavanine.

Day 12.

Using the multipronged device, patch or spot 45 independent random spore colonies from the 4-1 x 4-2 diploid onto YPD, SC-ura, SC-his, SC-met, and SC-arg+canavanine plates. Include as controls the 4-1 x 4-2 diploid and the parental haploids 4-1 and 4-2.

Day 15.

Meiotic recombination. Based on the growth of spotted spore colonies, determine the frequency of meiotic recombination between the *CAN1*, *URA3*, *HIS1*, and *HOM3* genes.

MATERIALS

Day 1.

1 YPD plate

Day 3.

1 YPD plate
1 SD+lys plate
Sterile loop or toothpicks
Sterile H_2O

Day 5.

1 YPD plate
2 Culture tubes containing 5 ml of YPD

Day 6.

Sterile microcentrifuge tubes
5 SC-arg+canavanine plates
5 SC+5-FOA plates
4 YPD plates
Liquid sporulation medium
Sterile H_2O
Glass spreader
70% Ethanol for sterilization

Day 9.

2 SC-arg+canavanine plates
2 SC+5-FOA plates
2 SC-met
2 YPD plates
Sterile toothpicks/dowels
Multipronged inoculating device
2 Sterile 96-well plates
Materials for breaking apart tetrads from Techniques and Protocols #9, Random Spore Analysis

Day 12.

1 SC-arg+canavanine plate
1 SC-ura plate
1 SC-met plate
1 SC-his plate
1 YPD plate
Sterile toothpicks/dowels
Multipronged inoculating device
1 Sterile 96-well plate

DATA

Group #

Spontaneous mutation (strain 4-2)

Medium	Dilution	# of Colonies	Freq. of mutation
5-FOA	10^{-1}		
	10^{-2}		
CAN	10^{-1}		
	10^{-2}		
YPD	10^{-4}		
	10^{-5}		

Loss of heterozygosity (strain 4-1 x 4-2)

Medium	Dilution	# of Colonies	Freq. of LOH
5-FOA	10^{-1}		
	10^{-2}		
	10^{-3}		
CAN	10^{-1}		
	10^{-2}		
	10^{-3}		
YPD	10^{-4}		
	10^{-5}		

5-FOA^R 4-1 x 4-2

#	Can	Met
1		
2		
3		
4		
5		
6		
7		
8		
9		
10		
11		
12		
13		
14		
15		
16		
17		
18		
19		
20		
21		
22		
4-1		
4-2		
4-1 x 4-2		

Can^R 4-1 x 4-2

#	5-FOA	Met
1		
2		
3		
4		
5		
6		
7		
8		
9		
10		
11		
12		
13		
14		
15		
16		
17		
18		
19		
20		
21		
22		
4-1		
4-2		
4-1 x 4-2		

Meiotic recombination (random spore analysis)

#	Can^R	Ura	His	Met
1				
2				
3				
4				
5				
6				
7				
8				
9				
10				
11				
12				
13				
14				
15				
16				
17				
18				
19				
20				
21				
22				
4-1				
4-2				
4-1x4-2				

#	Can^R	Ura	His	Met
23				
24				
25				
26				
27				
28				
29				
30				
31				
32				
33				
34				
35				
36				
37				
38				
39				
40				
41				
42				
43				
44				
45				

EXPERIMENT V

Transformation of Yeast

Saccharomyces cerevisiae is unique among eukaryotes in the ease with which it can be transformed with DNA, and the high frequency with which the introduced DNA undergoes homologous recombination with genomic DNA. There are several requirements for a successful transformation experiment: (1) a means of introducing DNA into cells; (2) a selectable marker on the introduced DNA with corresponding nonreverting mutations in the chromosome; and (3) vector systems that allow propagation of cloned DNA in both *E. coli* and yeast.

TRANSFORMATION METHODS

The original method for yeast transformation involved incubating spheroplasted cells with DNA in the presence of polyethylene glycol (PEG) and $CaCl_2$ (Hinnen et al. 1978). A more convenient and much more widely used method involves the treatment of cells with the alkali salt, lithium acetate (LiAc), followed by incubation with DNA and PEG (Ito et al. 1983). DNA can also be introduced by electroporation (Hashimoto et al. 1985; Becker and Guarente 1991), whereby a brief electrical pulse permeabilizes the cells to DNA; by agitation of cells with glass beads (Costanza and Fox 1988); by bombardment of cells with DNA-coated particles (currently the only way to transform mitochondria) (Fox et al.1988; Johnston et al. 1988); and by direct conjugation between bacterial and yeast cells (Heinemann and Sprague 1989). The method of choice depends on the purpose of the experiment, the number of strains to be transformed, and the desired number of transformants. The efficiency of transformation is often the most important parameter: if the goal is simply to put a plasmid into a given strain (only a few colonies needed), then any of the methods will work with almost any strain of yeast; if the goal is to get 10^6 transformants for screening a library, then the strain and transformation method must be carefully chosen. Spheroplasting, LiAc, and electroporation can all give high transformation frequency under optimal conditions.

SELECTABLE MARKERS

Although the frequency of transformation of yeast can be quite high, only a small fraction of the total number of cells in a transformation experiment become transformed. Therefore, it is essential to have reliable selectable markers to select for those cells that have become transformed. The most common selectable markers used in yeast transformation are those that complement a specific auxotrophy. For example, the yeast *LEU2* gene encodes β-isopropylmalate dehydrogenase and complements the leucine auxotrophy of a *leu2* mutant. Other commonly used selectable markers are *URA3*, mutations in which result in uracil auxotrophy; *HIS3*, mutations in which result in histidine auxotrophy; and *TRP1*, mutations in which result in tryptophan auxotrophy. Of equal importance to the selectable marker is the corresponding chromosomal mutation that causes the auxotrophy. This mutation should be completely recessive and nonreverting; for example, the *leu2-3,112* mutation is a double frameshift mutation that reverts with a very low frequency ($<10^{-10}$) and is completely complemented by the wild-type *LEU2* gene. More recently, dominant drug resistance has been used as a selectable marker in yeast, as it is used in bacterial transformation (Hadfield et al.1990). This has the potential advantage that the selectable marker lacks any homology with the yeast genome.

VECTOR SYSTEMS

Yeast cells that have taken up DNA during the transformation process can maintain that DNA, and thus become transformed, either by integration of the DNA into a chromosome or by autonomous replication. Integration into a chromosome takes place almost exclusively by homologous recombination in yeast. Once integrated, the transforming DNA is part of the chromosome and segregates in mitosis and meiosis with the same high fidelity as a chromosome. Plasmids used for integration have a yeast selectable marker, but no other yeast elements. Autonomous replication requires that the transforming DNA have a yeast origin of DNA replication. Originally called *ARS* (autonomously replicating sequence) elements, these can be either chromosomal DNA replication origins or the origin from the endogenous yeast 2μ plasmid. Because the yeast replication origin is a relatively simple and short DNA sequence, DNA from other organisms will occasionally be found to have yeast *ARS* activity. To transform yeast cells stably, transforming DNA must have either sufficient homology to the yeast genome to integrate, or carry an *ARS* element.

Plasmids with only a chromosomal *ARS* element (ARS plasmid) have a variable copy number and often fail to segregate to the daughter cell in a division (Murray and Szostak1983), resulting in a high rate of plasmid loss. Autonomously replicating plasmids may also have a centromere, or *CEN* element. A *CEN/ARS* plasmid (CEN plasmid) is more stable than a simple *ARS* plasmid because the centromere mediates the attachment of the plasmid to the mitotic spindle, ensuring segregation to both mother and daughter cells. Because of the high fidelity of segregation, the copy number is maintained at one or two plasmids per cell. CEN plasmids typically show a 2:2 (if the cell had one copy) or 4:0 (if the cell had at least two copies) segregation pattern in meiosis.

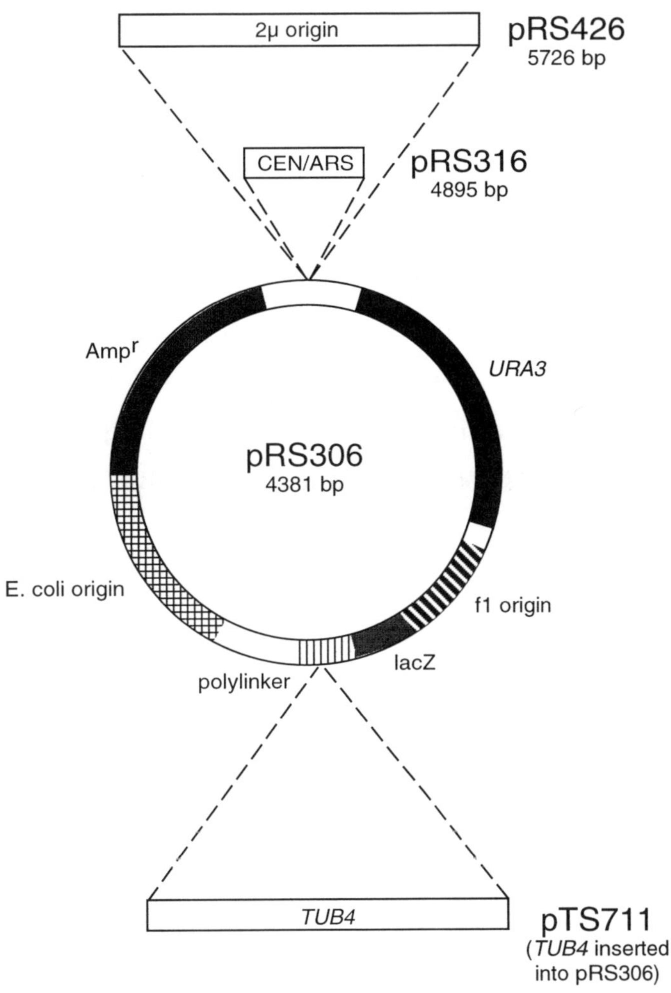

Figure 1 Plasmids used in this experiment.

An autonomously replicating plasmid that has the 2μ origin of replication (2μ-based plasmid) segregates in mitosis with about the same fidelity as a CEN plasmid, but is present at much higher copy number, typically 20–50 copies per cell. The high fidelity of segregation of 2μ plasmids depends on the presence of the endogenous 2μ plasmid. If a strain lacks the endogenous 2μ plasmid, then introduced 2μ-based plasmids will segregate as ARS plasmids (Murray and Szostak 1983).

Many older yeast plasmids were named systematically. In this system, integrating plasmids were designated YIp; ARS plasmids, YRp; CEN plasmids, YCp; and 2μ plasmids, YEp (E for episome). YRp plasmids are rarely used because of their extreme instability.

In addition to the yeast-specific elements, all standard yeast vectors also have a bacterial origin of replication and a bacterial selectable marker, usually ampicillin resistance. Older vectors (YIp5, YEp24, YCp50) are usually based on a pBR322 backbone. More recent plasmids are often based on plasmid backbones that replicate to higher copy number in bacteria, have useful polylinker sequences, and have single stranded phage origins for the isolation of DNA for sequencing. A well-designed set of yeast vectors was described by Sikorski and Hieter (1989), and are used in this experiment. Figure 1 illustrates the plasmids to be used, all based on pRS306, an integrating vector with *URA3* as the selectable marker: pRS316 is pRS306 with an inserted *CEN/ARS* element, and pRS426 is a derivative of pRS 306 with an inserted 2μ origin (the orientation of the polylinker is also reversed). pTS711 is pRS306 with a genomic fragment including the yeast *TUB4* gene inserted in the polylinker.

EXPERIMENT V(a)

Integration

In most cases, integration of a circular plasmid into the yeast genome occurs by a single crossover and yields a direct repeat of the yeast sequence on the plasmid, as shown in Figure 2. Note that the entire plasmid is integrated, including the bacterial sequences and the yeast selectable marker—these now serve as a physical and phenotypic marker for the site of integration. The plasmid used in this experiment is pTS711, which has two regions of homology with the yeast genome: the *URA3* gene that is part of the parent vector, pRS306, and the *TUB4* gene that was inserted into the polylinker of pRS306.

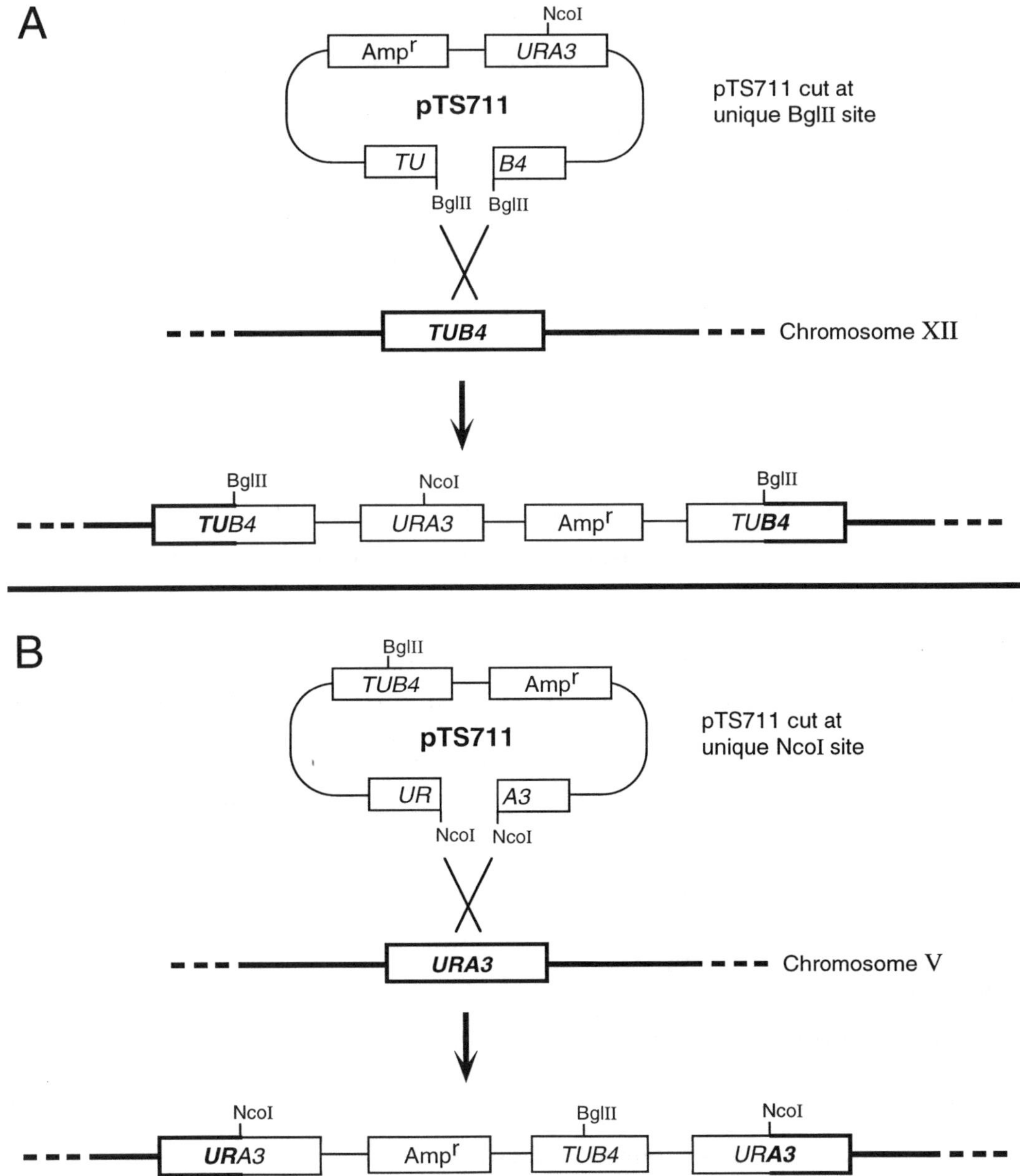

Figure 2 Integration of plasmid pTS711 at either the *TUB4* or *URA3* locus.

TUB4 encodes γ-tubulin, an essential protein involved in spindle pole body function (Marschall et al. 1996).

pTS711 cannot replicate in yeast because it lacks an *ARS* element and can, therefore, only transform yeast by integration. Integration can occur at either the *URA3* locus (actually the *ura3-52* locus in the strain used here) or the *TUB4* locus by homologous recombination (Fig. 2). Cutting a plasmid within a region of homology with the genome greatly increases the specificity and frequency of recombination at that site (Orr-Weaver et al. 1981); the free ends are highly recombinogenic and increase the efficiency of integration, and thus transformation, by approximately 100-fold. In this experiment, uncut pTS711 should transform strain 5-1 with a very low efficiency and integrate at either *ura3* or *TUB4*. Cutting pTS711 at the unique *Nco*I site in *URA3* should result in nearly 100% integration at *ura3*, and cutting pTS711 at the unique *Bgl*II site in *TUB4* should result in nearly 100% integration at *TUB4*.

Integration at a locus can be verified both genetically and physically. In this experiment, transformants will be crossed to tester strains to determine the site of integration. Integration at *TUB4* will be tested by crossing to a strain with a $tub4^{ts}$ mutation. If the plasmid integrates at the *TUB4* locus, then the plasmid-borne *URA3* gene will be genetically linked to the *TUB4* locus; thus every tetrad should be a PD with two Ura^+ Ts^+ and two Ura^- Ts^- spores. Integration at *ura3* will be tested by crossing to a *URA3* strain. If the plasmid integrates at the *ura3* locus, then the plasmid-borne *URA3* gene will be genetically linked to the *ura3* locus; thus every tetrad should be a PD with four Ura^+ spores. As an exercise, predict the outcome of the cross if the plasmid had integrated at *TUB4* instead of *ura3*.

EXPERIMENT V(b)
Replication and Stability

Strain 5-1 will also be transformed with the autonomously replicating plasmids pRS316 and pRS426. The frequency of transformation will be higher for these plasmids than for the integrating plasmid. The stability of these plasmids in mitosis and meiosis will be examined.

STRAINS

5-1	TSY623	*MATα ade2-101 his3-Δ200 leu2-3,112 ura3-52*
5-2	TSY502	*MAT***a** *his3-Δ200 leu2-3,112 lys2-801 ura3-52 tub4-34*
5-3	TSY807	*MAT***a** *his3-Δ200 leu2-3,112 lys2-801 ura3-52*
5-4	TSY808	*MAT***a** *lys2-801*

PLASMIDS

pTS711	*TUB4 URA3*
pRS316	*CEN URA3*
pRS426	2μ *URA3*

PROCEDURE

Day 1.

Subclone strain 5-1 on a YPD plate. Incubate at 30°C.

Day 3.

In the morning, inoculate 5 ml of YPD with a single colony of strain 5-1 to make a preculture. Incubate at 30°C with agitation. In the evening, determine the cell concentration using a hemocytometer. Assuming an approximate generation time of 100 minutes, calculate the volume of preculture to be added to 100 ml of YPD in order for the culture to reach a concentration of 2×10^7 cells/ml by 10 a.m. on Day 4.

Day 4.

Harvest the cells by centrifugation at 2000 rpm in a clinical centrifuge and follow the protocol for LiAc transformation given in Techniques and Protocols #1, High-efficiency Transformation of Yeast. Transform 5-1 under the conditions listed below.

Expt.	Recipient	DNA	Digestion	Selective medium
A	5-1	pTS711	none	SC-ura
	5-1	pTS711	*Nco*I	SC-ura
	5-1	pTS711	*Bgl*II	SC-ura
B	5-1	pRS316	none	SC-ura
	5-1	pRS426	none	SC-ura
Control	5-1	none	none	SC-ura

Day 7.

(i) In order to analyze the transformants genetically, they will crossed to tester strains. Carefully pick two transformant colonies from each of the transformation plates with sterile toothpicks or a loop and mix them with the tester strains in a small patch on a YPD plate. Cross the strains in the combinations shown below, and incubate overnight at 30°C (for crosses involving 5-2, which contains a ts mutation, incubate at room temperature).

Transformant	Tester	Selective Medium
Integration location		
5-1+pTS711(*Nco*I)	5-4	SD
5-1+pTS711(*Nco*I)	5-2	SD+his+leu
5-1+pTS711(*Bgl*II)	5-4	SD
5-1+pTS711(*Bgl*II)	5-2	SD+his+leu
Meiotic Stability		
5-1+pTS711(*Bgl*II)	5-3	SD+his+leu
5-1+pRS316	5-3	SD+his+leu
5-1+pRS426	5-3	SD+his+leu

(ii) The assays for mitotic stability require that transformants be purified away from the untransformed cells on the transformation plates. Purify two transformants each from the *Bgl*II-cut integrating transformation (Expt. V(a) on Day 4) and the autonomously replicating plasmid transformations (Expt. V(b) on Day 4) by streaking onto SC-ura agar. Incubate at 30°C.

Day 8.

Replica-plate the YPD plates containing the crosses onto the appropriate selective medium. Only diploid cells formed by mating should be able to grow in the patches where the cells were mixed. Incubate at 30°C.

Day 9.

i) In preparation for sporulation, patch the diploids growing on the selective plates to a YPD plate and incubate for 1 day at room temperature.

(ii) Inoculate one 5-ml YPD culture with a toothpickful of cells from one colony of each purified transformant from the SC-ura plates. Place the tubes on the roller drum overnight at 30°C.

Day 10.

(i) Transfer the diploids from YPD to sporulation medium (both liquid and plate to compare efficiency of sporulation) and incubate several days at room temperature.

(ii) Make serial dilutions of the overnight 5 ml YPD cultures in sterile H_2O and spread aliquots on YPD plates. Try to get 100–300 colonies/plate. Incubate at 30°C.

Day 12.

(i) Begin dissecting tetrads from the crosses involving the transformants. Try to dissect ten tetrads from each cross. Incubate tetrad plates at 30°C, except for those involving 5-2, which should be incubated at room temperature.

(ii) Replica-plate YPD plates with the appropriate number of colonies to SC plates and SC-ura plates. Incubate at 30°C.

Day 13–14.

(i) As tetrads grow up on the dissection plates, replica-plate the crosses to strains 5-3 and 5-4 onto SC-ura and YPD plates at 30°C to determine segregation of *URA3*. Replica-plate crosses to 5-2 onto YPD plates at both room temperature and 37°C to determine segregation of *TUB4*.

(ii) Score the fraction of cells that remained Ura^+ after nonselective growth in YPD.

MATERIALS

Day 1. 1 YPD plate

Day 3. 1 Culture tube containing 5 ml of YPD
Erlenmeyer flask containing 100 ml of YPD

Day 4. Materials for Techniques and Protocols #1, High-efficiency Transformation of Yeast
7 SC-ura plates

Day 7. 3 YPD plates
3 SC-ura plate

Day 8. 1 SD plate
2 SD+his+leu plates

Day 9. 2 YPD plates
3 Culture tubes, each containing 5 ml of YPD

Day 10. 15 Culture tubes, each containing 5 ml of sterile H_2O
12 YPD plates
7 Culture tubes, each containing 2 ml of sporulation medium
2 sporulation plates

Day 12. 10 YPD plates for tetrad dissection
6 SC plates
6 SC-ura plates

Day 13. 7 SC-ura plates
7 YPD plates

REFERENCES

Becker, D. M. and L. Guarente. 1991. High-efficiency transformation of yeast by electroporation. *Methods Enymol.* **194:** 182–187.

Costanza, M.C. and T.D. Fox. 1988. Transformation of yeast by agitation with glass beads. *Genetics* **120:** 667–670.

Fox, T.D., J.C. Sanford, and T.W. McMullin. 1988. Plasmids can stably transform yeast mitochondria lacking endogenous mtDNA. *Proc. Nat. Acad. Sci.* **85:** 7288–7292.

Hadfield, C., B.E. Jordan, R.C. Mount, G.H.J. Pretorius., and E. Burak. 1990. G418-resistance as a dominant marker and reporter for gene expression in *Saccharomyces cerevisiae*. *Curr. Genet.* **18:** 303–314.

Hashimoto, H., H. Morikawa, Y. Yamada, and A. Kimura. 1985. A novel method for transformation of intact yeast cells by electroinjection of plasmid DNA. *Appl. Microbiol. Biotechnol.* **21:** 336–339.

Heinemann, J.A. and G.F. Sprague. 1989. Bacterial conjugative plasmids mobilize DNA transfer between bacteria and yeast. *Nature* **340:** 205–209.

Hinnen, A., J.B. Hicks, and G.R. Fink. 1978. Transformation in yeast. *Proc. Natl. Acad. Sci.* **75:** 1929–1933.

Ito, H., Y. Fukuda, K. Murata, and A. Kimura. 1983. Transformation of intact yeast cells treated with alkali cations. *J. Bacteriol.* **153:** 163–168.

Johnston, S.A., P. Anziano, K. Shark, J.C. Sanford, and R.A. Butow. 1988. Transformation of yeast mitochondria by bombardment of cells with microprojectiles. *Science* **240:** 1538–1541.

Marschall, L., R. Jeng, J. Mulholland, and T. Stearns. 1996. Analysis of Tub4p, a yeast γ-tubulin-like protein: Implications for microtubule organizing center function. *J. Cell Biol.* **134:** 443–454.

Murray, A.W. and J.W. Szostak. 1983. Pedigree analysis of plasmid segregation in yeast. *Cell* **34,** 961–970.

Orr-Weaver, T., J. Szostak, and R. Rothstein. 1981. Yeast transformation: A model system for the study of recombination. *Proc. Nat. Acad. Sci.* **78:** 6354–6358.

Sikorski, R.S. and P. Hieter. 1989. A system of shuttle vectors and yeast host strains designed for efficient manipulation of DNA in *Saccharomyces cerevisiae*. *Genetics* **122:** 19–27.

Experiment VI

Cytoduction and Karyogamy

The sexual cycle in haploid strains of *Saccharomyces cerevisiae* occurs by an orderly progression of sequential events: cell fusion to form zygotes, nuclear fusion (karyogamy) to form diploids, and meiosis to form haploids. In normal strains of *S. cerevisiae*, nuclear fusion follows immediately after cell fusion, with no intervening cell or nuclear divisions. Once the nuclei have fused, the zygote gives rise to vegetative diploid cells by mitotic budding.

A number of genes, including *KAR1*, are required for efficient karyogamy following cell conjugation. Mutations in these genes cause zygotes to produce buds of haploid cytoductants, rather than true diploid progeny, at a high frequency (see Berlin et al. 1991). (Cytoductants are exconjugants that contain the nucleus of one parent and cytoplasmic markers from both parents.) Thus, such mutants allow the transmission of non-nuclear genetic elements from one haploid strain to another without the necessity of a meiotic cross.

Interestingly, it has been found that in exceptional cytoductants from karyogamy mutants, nuclear material can be exchanged between nonfused nuclei (Dutcher 1981). In these cases, a plasmid or a single whole chromosome is transferred from one nucleus to another. The transfer of a chromosome typically creates a disomic strain; there is one pair of homologous chromosomes in an otherwise haploid nucleus.

In this experiment, cytoductants will be created by forming zygotes derived from two strains of opposite mating type, one of which contains a *kar1* mutation. In order to identify true haploid cytoductants, one of the mating partners will contain a recessive drug resistance marker (e.g., *cyh2*^{r} for cycloheximide resistance; *can1*^{r} for canavanine resistance) while the other will be sensitive to the drug (e.g., *CYH2*^{S}, *CAN1*^{S}). This permits a simple way to select against diploid progeny.

In order to demonstrate transmission of a cytoplasmic element, a strain with defective mitochondria will be created. Specifically, ρ^- mutants (petites), which are defective in mitochondrial function, will be made by exposing cells to ethidium bromide in order to induce loss of mitochondrial DNA. A simple method to distinguish between ρ^- (mutant) and ρ^+ (normal) cells is to plate them onto a YPDG plate and incubate for 3–4 days at 30°C. At this time, the ρ^+ colonies are several times larger than the ρ^- colonies because ρ^- cells cannot utilize glycerol for growth, and glucose is in growth-limiting amounts in the media.

Lastly, a nuclear plasmid will be exchanged between karyogamy-deficient nuclei. A haploid strain carrying a nuclear plasmid will be made and conjugated with a second strain that lacks the plasmid. Exceptional cytoductants will be identified that have the nuclear content of the second strain yet also contain the plasmid from the first strain.

STRAINS

6-1	DG21	*MAT***a** *kar1 aro2 ade3 ura3*
6-2	DDM30B	*MATα ade2 can1 ura3Δ trp1Δ leu2 ade5 cyh2^R lys5*
6-3	PT1	*MAT***a** *ile hom3 can^R*
6-4	PT2	*MATα ile hom3 can^R*

PLASMID

pRS316 — An *E. coli* - yeast shuttle vector that contains an ARS and *CEN6* to keep the plasmid at low copy when selection for its *URA3* gene is applied. This is a handy vector developed by Sikorski and Hieter (1989).

PROCEDURE

Safety Note

Ethidium bromide is a powerful mutagen and is moderately toxic. Wear gloves when working with solutions that contain this dye. Consult the local institutional safety officer for specific handling and disposal procedures.

Day 1.

Streak strains 6-1, 6-2, 6-3, and 6-4 for single colonies onto YPD plates. Incubate for 2 days at 30°C.

Day 3.

(i) Pick one or two large colonies of strain 6-1 with a toothpick from the YPD plate and place in a sterile 1.5-ml Eppendorf tube. Transform these 6-1 cells with plasmid pRS316 (see Techniques and Protocols #2, "Lazy Bones" Plasmid Transformation of

Yeast Colonies). Make certain to perform a "control" transformation in which no plasmid DNA is added. Incubate cells overnight on your benchtop.

(ii) In the evening, inoculate two 5-ml cultures of HC with strain 6-2. To one of the two cultures, add ethidium bromide to a final concentration of 10 μg/ml. Put both cultures on a shaker or roller drum at 30°C in the dark. (Wrap the tube in foil if necessary).

(iii) Resuspend a large colony of strain 6-3 from the YPD plate into 1 ml of sterile H_2O. Plate 0.2 ml of the suspension onto a fresh YPD plate; spread uniformly. This will create a "lawn" of cells for use later. Perform the same task for strain 6-4. Incubate for 2 days at 30°C.

Day 4.

(i) Finish transforming the strain 6-1 cells with plasmid pRS316, making certain to heat shock the cells for 15 minutes at 42°C (see Techniques and Protocols #2, "Lazy Bones" Plasmid Transformation of Yeast Colonies). Select for the plasmid by plating onto HC-ura media, and incubate for 2 days at 30°C.

(ii) Determine the cell density of each of the two HC media cultures of strain 6-2 (one with ethidium bromide, one without) using a hemocytometer (see Appendix G). Make a dilution of each culture to give ~500–1000 cells/ml. (The dilution can be made with sterile H_2O.) Spread 200 μl of each of the two dilutions onto YPDG plates (make certain to mark the plates so as to distinguish between the ethidium bromide and untreated cells), and incubate for 3 days at 30°C.

Day 5.

Remove the YPD plates of strains 6-3 and 6-4 from the 30°C incubator and store at 4°C. These will be used on Day 14.

Day 6.

Pick three of the strain 6-1 Ura^+ transformants and patch onto an HC-ura plate. Incubate for 2 days at 30°C. Save the transformation plate by storing at 4°C.

Day 7.

Record the number of small (ρ^-) and large (ρ^+) colonies on each of the two YPDG plates and calculate the frequency at which ρ^- cells arose in strain 6-2 under the two growth conditions. Pick one ρ^- colony from each plate (pre-grown with or without ethidium bromide) and patch onto a YPD plate. Incubate for 2 days at 30°C.

Day 8.

Divide a single YPD plate into thirds, pick two dollops of cells from one of the 6-1 Ura^+ transformant patches, and patch each in a distinct third of a single YPD plate. To one of the 6-1 Ura^+ patches, add an equivalent-size dollop of strain 6-2 cells (from the plate inoculated on Day 1), and mix thoroughly on the plate. Also patch a dollop of 6-2 cells to the remaining third of the plate without mixing. Incubate the plate for 6–8 hours at 30°C. At the end of the incubation time, scrape all the cells within the mixed patch of 6-1 Ura^+ and 6-2 cells and streak for single colonies onto an HC-ura+cyh plate; repeat this procedure for the 6-2 and 6-1 Ura^+ patches. (These serve as the negative controls for the experiment.) Incubate for 4 days at 30°C.

Day 9.

Pick a dollop of cells from each of the two ρ^- 6-2 patches (from Day 7) and mix with 6-1 cells (not the ones transformed with the URA3 plasmid) as described above, only this time divide the YPD plate into fourths. Make certain to include negative controls of 6-1 and ρ^- 6-2 cells that are not mixed together. Incubate for 6–8 hours at 30°C. At the end of the incubation, scrape the cells and streak for single colonies onto YPG+cyh plates. Incubate for 4 days at 30°C.

Day 12.

From the cells plated on Day 8, note the growth of cells on the HC-ura+cyh plates, comparing the mated cells with each of the unmixed controls. Pick three colonies from the mixed cells on the HC-ura+cyh plate and make a line ~2 cm long of each isolate on an HC-ura plate. Separate each line by ~0.5 cm. Also place one of the original 6-1 Ura^+ transformants on the same plate as a control. Incubate for 2 days at 30°C.

Day 13.

Note the growth of cells on the YPG+cyh plates from Day 9; compare the mated cells with each of the unmixed controls. Pick three colonies from the mixed cells from the YPG+cyh plate and patch to an YPD plate. Also patch strain 6-1 and the 6-2 ρ^- mutant onto the same plate as controls. Incubate for 1 day at 30°C.

Day 14.

(i) Replica-plate the patched cells from Day 12 and Day 13 onto the following set of plates:

HC-ade
HC-trp
HC-leu
HC-ura
HC-lys
HC+can-arg
YPD+cyh
YPG
YPD

(ii) Cross-stamp each of the patches replica-plated to YPD with the lawns of strains 6-3 and 6-4. Incubate all plates at 30°C.

Day 15.

(i) Replica-plate each cross-stamped YPD plate to an SD plate.

Day 16.

Score all the plates, including the SD plate, to determine whether the isolates are haploid.

MATERIALS

Day 1.	4 YPD plates Sterile H_2O Sterile Eppendorf tubes
Day 3.	Material for Techniques and Protocols #2, "Lazy Bones" Plasmid Transformation of Yeast Colonies 1 μg of pRS316 plasmid DNA 2 Culture tubes, each containing 5 ml of HC media 1 mg/ml of Ethidium bromide stock solution (made with sterile H_2O) Sterile H_2O
Day 4.	2 HC-ura plates 2 YPDG plates
Day 6.	1 HC-ura plate

Day 7. 1 YPD plate

Day 8. 1 YPD plate
3 HC-ura+cyh plates

Day 9. 1 YPD plate
4 YPG+cyh plates

Day 12. 1 HC-ura plate

Day 13. 1 YPD plate

Day 14. 2 of each of the following plates:
HC-ade
HC-trp
HC-leu
HC-ura
HC-lys
HC+can-arg
YPD+cyh
YPG
YPD
Sterile tongue depressors

Day 15. 2 SD plates

This experiment was heavily influenced by and partially taken from earlier versions of this manual written by F. Sherman, G. Fink, and J. Hicks.

REFERENCES

Berlin, V., J.A. Brill, J. Trueheart, J.D. Boeke, and G.R. Fink. 1991. Genetic screens and selections for cell and nuclear fusion mutants. *Methods Enzymol.* **194:** 774.

Dutcher, S.K. 1981. Internuclear transfer of genetic information in kar1-1/KAR1 heterokaryons in *Saccharomyces cerevisiae*. *Mol. Cell. Biol.* **1:** 245.

Sikorski, R.S. and P. Hieter. 1989. A system of shuttle vectors and yeast host strains designed for efficient manipulation of DNA in *Saccharomyces cerevisiae*. *Genetics* **122:** 19–27.

Experiment VII

Gene Replacement

One of the most powerful and important techniques available for studies in yeast is gene replacement. This technique allows the replacement of a gene at its normal chromosomal location with an allele of that gene created *in vitro*, such that the only genetic difference between the initial strain and the final strain is that particular allele. Using this method, phenotypes conferred by null mutations or any other types of mutations made in a cloned gene can be analyzed. In theory and in practice, a cloned yeast gene can be changed and then recombined into the genome, precisely replacing the wild-type allele.

Determination of the null phenotype for a gene is an essential step in understanding the function of that gene. First, it reveals whether the gene is essential for growth. Second, if the gene is not essential for growth, it allows study of strains completely lacking that particular function. To examine the phenotype conferred by a null mutation, gene replacement is generally done in diploid strains, in case the null mutant is inviable as a haploid. The inviability will be observed after tetrad analysis of a sporulated culture. If the gene is essential, viability will segregate 2:2 in the tetrads. If the gene is not essential, all four spores will be viable and two will be null mutants for that gene.

Experiment VII(a)

One-step Gene Disruption

Traditionally, a one-step gene replacement (Rothstein 1983) has been done by transformation with a restriction fragment that contains the mutant allele and has ends homologous to the locus where integration is desired. However, in order to create the mutant allele fragment, the gene of interest had to be cloned, and then using convenient restriction sites, the gene, or a portion of it, was replaced by a DNA fragment encoding a selectable marker in yeast. Because of the inherent limitations of these manipulations, it is frequently difficult to create a true null allele, in which the entire coding sequence of a gene is precisely removed. However, it is now much easier to make null mutations in yeast using the PCR-mediated gene disruption method (Baudin et al. 1993; Lorenz et al. 1995).

PCR-mediated gene disruption is based on the fact that homologous recombination in yeast is very efficient with linear DNA fragments, and that only ~40 bp of homology is required for rather efficient recombination. The availability of the entire *S. cerevisiae* genome sequence in combination with PCR-mediated gene disruption has made it possible to create a null mutation of any gene without ever having to clone it. Furthermore, a series of plasmids and strains have been created that increase the efficacy of this technique (Brachmann et al. 1997). The plasmids contain common yeast selectable marker genes cloned into a conserved site. This permits one set of PCR primers to be used to disrupt a gene with a choice of markers (see Fig. 1). The yeast strains have the common auxotrophic markers completely deleted from the genome, which eliminates the unwanted background of gene conversion events that can happen between the selectable markers and their mutant alleles in the chromosome.

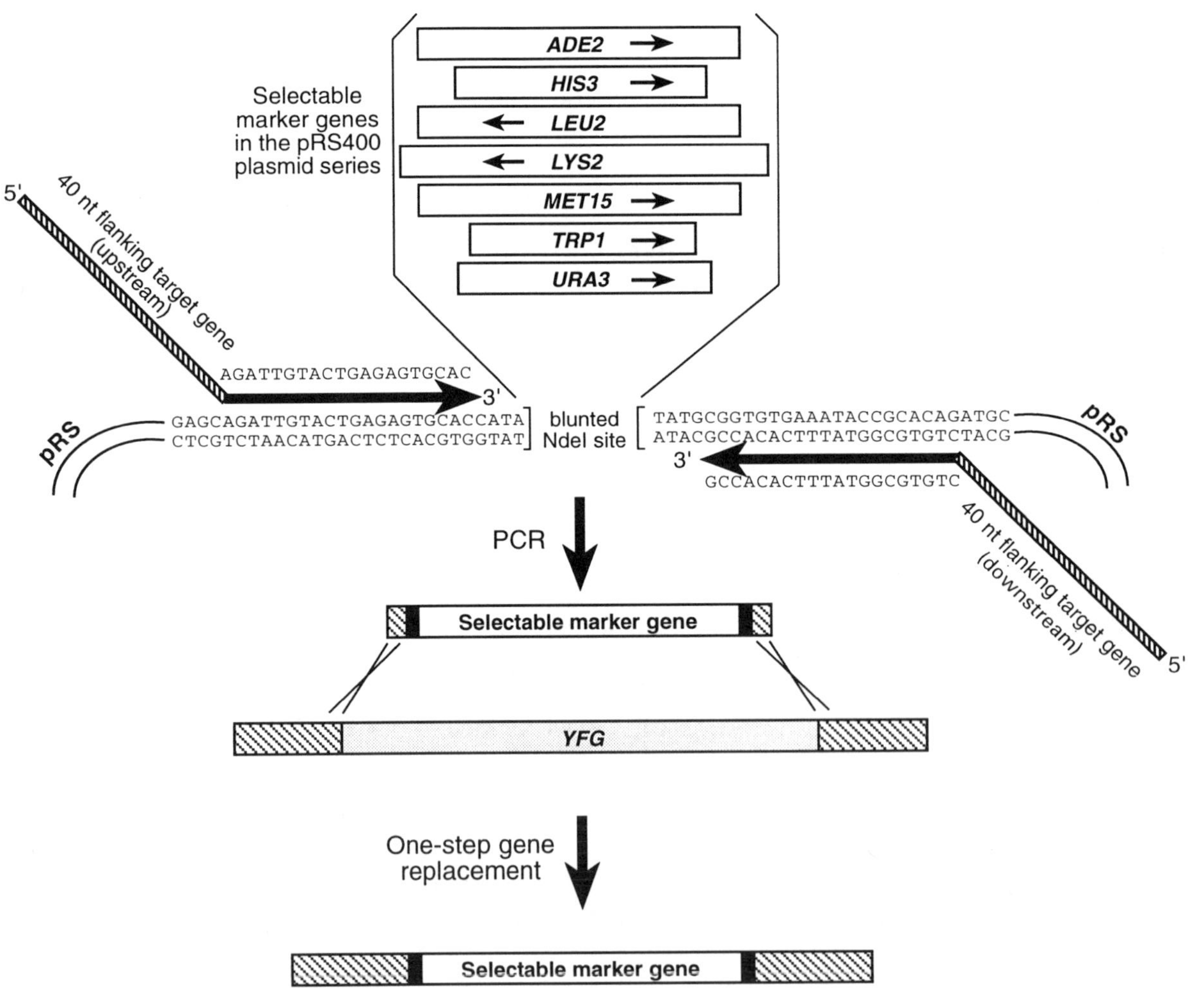

Figure 1 One-step gene replacement.

Often gene disruptions are carried out to test the phenotype of a null allele of a gene of interest. In this experiment, we will disrupt the gene for a protease in order to make a strain that will be useful for biochemical experiments. One of the difficulties in doing biochemistry on yeast cell extracts is the abundance of proteases contained within the vacuole that are released when cells are broken open. A truly protease deficient strain requires disruption of each of the different genes that encode vacuolar proteases. Fortunately, there is a way to eliminate most of the proteases in a single step. Vacuolar proteases are synthesized as inactive pro-enzymes that become active in the vacuole on proteolytic removal of the pro-peptide. The major processing protease, proteinase A, is encoded by the *PEP4* gene. Disruption of *PEP4* prevents all of the other proteases from becoming active. A complication of this procedure is that proteinase B can also cleave pro-peptides and activate vacuolar proteases. In the absence of proteinase A, active proteinase B can continue to process vacuolar proteases for several cell divisions, but the self-activation of proteinase B is not sustained indefinitely. Once the activity of proteinase B falls below a threshold, the cell becomes irreversibly protease negative. After we disrupt *PEP4*, we will streak out the disruption strain for single colonies several times to dilute proteinase B and to obtain a protease negative strain.

A different PCR method will be used to screen transformed colonies for a disrupted *PEP4* gene. In this procedure, screening will be done on colonies directly, without the need of purifying DNA from yeast. Transformants that are genotypically *pep4* will then be screened, using a plate assay, for Carboxypeptidase Y activity. This peptidase is a vacuolar enzyme that requires proteinases A and B for activation.

STRAIN

7-1 BY4732 *MATα his3Δ200 met15Δ0 trp1Δ63 ura3Δ0*

PLASMID

pRS303 An *E. coli* vector carrying the yeast *HIS3* gene that can be used as a template for PCR-mediated gene disruption.

PROCEDURE

Day 1.

Streak strain 7-1 on a YPD plate. Incubate at 30°C.

Day 3.

Pick a robust colony of strain 7-1 and use it to inoculate a 5-ml YPD culture. Incubate overnight at 30°C.

Day 4.

Determine the cell density of the 5-ml strain 7-1 culture using the hemocytometer (see Appendix G). Dilute cells to 5 x 10^6 cells/ml in an Erlenmeyer flask containing 50 ml of YPD and grow culture for two additional divisions at 30°C (about 4 hours).

Harvest the cells and transform them, following Techniques and Protocols #1, High-efficiency Transformation of Yeast, with ~1 µg of the *pep4::HIS3* PCR-generated DNA (a stock will be provided that was made via Techniques and Protocols #13, PCR Protocol for PCR-mediated Gene Disruption, using RS303 as a vector template, and the two primers:

5′PEP4DEL

TGGTCAGCGCCAACCAAGTTGCTGCAAAAGTCCACAAGGC*AGATTGTACTGAGAGTGCAC*

3′PEP4DEL

AATCGTAAATAGAATAGTATTTACGCAAGAAGGCATCACC*CTGTGCGGTATTTCACACCG*). Also be sure to carry out a transformation with no DNA, as a negative control. Plate each transformation onto one HC-his plate. Incubate for 3 days at 30°C.

Day 7.

Pick 11 transformants and streak for single colonies on HC-his plates. Use only three plates and streak 3–4 transformants per plate. Incubate for 2 days at 30°C.

Day 9.

(i) Pick one colony from each of the 11 transformants that were streaked on HC-his (from Day 7) and determine whether the *PEP4* gene has been disrupted using Techniques and Protocols #14, Yeast Colony PCR Protocol. Also make certain to include a sample of strain 7-1 cells as a control. Amplify the DNA by using the two oligos:

5′PEP4 GCTTGAAAGCATTATTGCCATTGGCC

3′PEP4 GGCCAAACCAACCGCATTGTTGCCC

(ii) While the PCR amplification is being carried out, prepare a 1.5% agarose gel.

Check the PCR products from each colony by gel electrophoresis. Add 3 µl of agarose gel loading buffer to each PCR sample and load all of each sample on the gel. Make certain to include a DNA size marker. The wild-type *PEP4* gene should produce an ~1.2 kbp band, whereas the *pep4::HIS3* allele has a predicted size of ~1.4 kbp.

(iii) Streak out three of the *pep4::HIS3* transformants as well as strain 7-1 onto YEPD plates. Streak two strains per plate. Incubate for 3 days at 30°C.

Day 12.

Assay the colonies on the YEPD plates using the APE protease plate assay as in Techniques and Protocols #8, Plate Assay for Carboxypeptidase Y. Note any differences in color development between colonies.

MATERIALS

Day 1. 1 YPD plate

Day 3. 1 Culture tube containing 5 ml of YPD

Day 4. Materials for Techniques and Protocols #1, High-efficiency Transformation of Yeast
Erlenmeyer flask containing 50 ml of YPD
2 HC-his plates

Day 7. 3 HC-his plates

Day 9. Materials for Techniques and Protocols #14, Yeast Colony PCR Protocol
~120 pmoles of each DNA oligo:
5′PEP4 GCTTGAAAGCATTATTGCCATTGGCC
3′PEP4 GGCCAAACCAACCGCATTGTTGCCC
Materials and equipment for a 1.5% agarose gel
2 YPD plates

Day 12. Materials for Techniques and Protocols #8, Plate Assay for Carboxypeptidase Y

EXPERIMENT VII(b)

Two-step Gene Replacement

The second type of gene replacement commonly used is a two-step method (Scherer and Davis 1979) in which one first integrates a plasmid that contains the mutant allele. When the integration occurs at the locus of interest, it will result in a duplication of the region—one duplicate will be wild-type and the other mutant—with the plasmid sequences in between. Homologous crossovers between the copies from the duplication will result in excision of the plasmid and loss of one of the two copies of the duplicated region. Depending on the exact location of the crossover, the copy left behind will contain either the mutant or wild-type form of the gene (Fig. 2). These can be distinguished by phenotype and/or by Southern hybridization analysis. Excision of the plasmid is done most conveniently if the selectable marker on the plasmid is URA3; one can simply select for those that have lost the plasmid using 5-fluoro-orotic acid (5-FOA) plates.

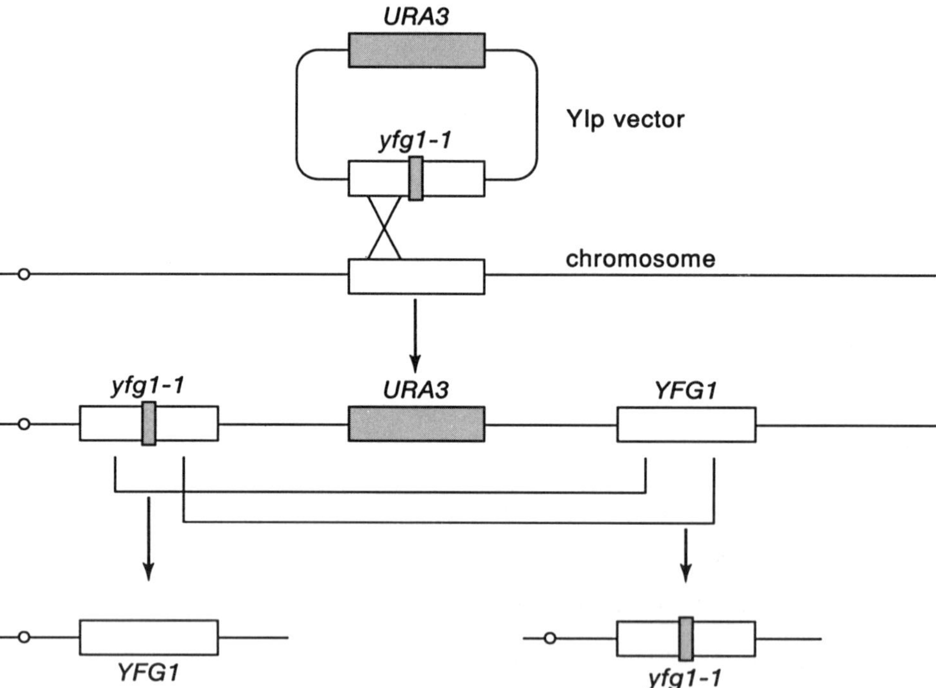

Figure 2 Two-step gene replacement. Integration of a plasmid by a single crossover results in a duplication of the *YFG1* locus. One copy contains the wild-type *YFG1* gene and the other copy contains the *yfg1-1* mutation. Strains that have excised the plasmid can be selected using 5-FOA medium. Depending on the location of the crossover, the excision will leave behind either *YFG1* or *yfg1-1*.

Two-step gene replacement is the method of choice under two conditions:

1. If the desired mutation is not associated with a selectable phenotype, direct selection by the one-step method is not possible. This is the case for gene replacement with a temperature sensitive allele of a particular gene. Using the two-step method, one can screen recombinants that have looped out the plasmid for those containing the mutant allele.
2. Using the two-step method, one can generate both the wild-type and mutant strains from the same initial transformant strain. This provides strains that are isogenic except in the gene of interest. A two-step gene replacement protocol will be used in Experiment IX to switch mating-type alleles at the *MAT* locus.

EXPERIMENT VII(c)

Plasmid Shuffling

In addition to analyzing phenotypes through gene replacement, it is possible to screen for mutant phenotypes when the gene of interest is on an autonomous plasmid, particularly when attempting to identify conditional lethal alleles of an essential gene. Using the plasmid-shuffle technique (Boeke et al. 1987), a plasmid containing the gene of interest can be mutagenized, the pool of mutagenized plasmids transformed into yeast, and then screened for a conditional mutant phenotype in the null background (Fig. 3). This is done by transforming a strain that already contains (1) the wild-type gene of interest on an autonomous plasmid with the *URA3* gene as its selectable marker and (2) a null allele of the gene of interest in the genome. In such a strain, viability is dependent on the function encoded on the plasmid. After transformation with the mutagenized pool of plasmids (using a selectable marker other than *URA3),* the phenotype of the potential mutant is screened for by replica plating onto 5-FOA plates under different conditions (e.g. temperature, media conditions). Only cells that have lost the wild-type gene on the *URA3* plasmid can grow on the 5-FOA plates, and the phenotype conferred by the mutant plasmid will be expressed.

While it has been known for some time that the DNA in eukaryotic chromosomes is intimately associated with histone proteins to form chromatin, the biological role of histones has only recently begun to be dissected. As is typical for most eukaryotes,

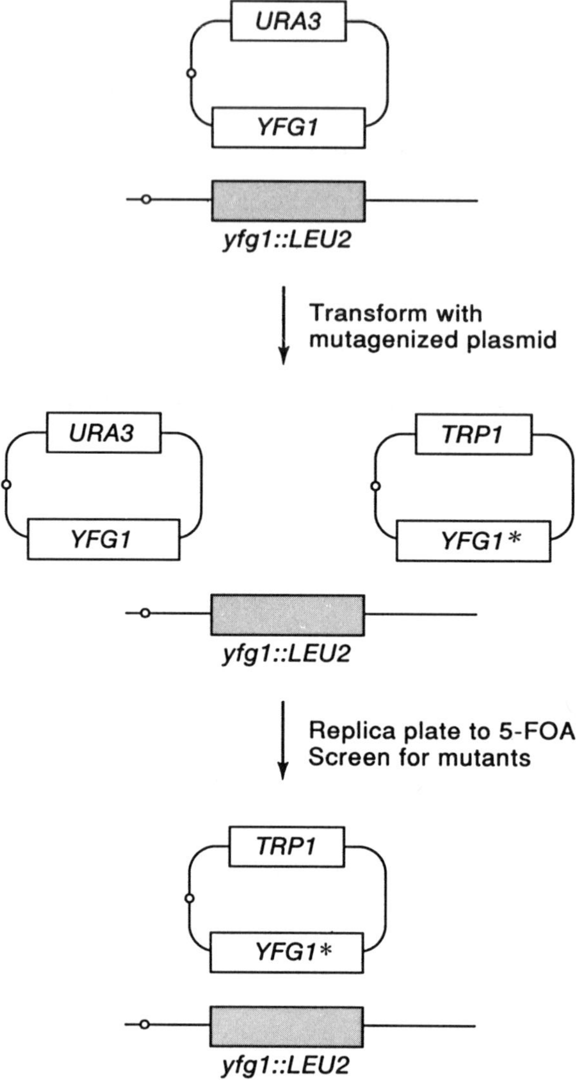

Figure 3 Plasmid shuffling.

there are four core histones in yeast chromatin, H2A, H2B, H3 and H4. Each histone protein has two unlinked loci within the genome (e.g. histone H3 is encoded by *HHT1* and *HHT2*), either of which is sufficient for viability. Curiously, each gene encoding histone H3 is very tightly linked to a gene for histone H4 (e.g. *HHT1* is adjacent to *HHF1* on the chromosome). The same situation is also true for histone H2A and H2B. In this experiment, we will take advantage of this situation and isolate two different types of mutations in the linked genes *HHT2* and *HHF2*. Alleles of histone genes have been isolated that affect activation of genes, such as those required in galactose metabolism (Durrin et al. 1991; Hirschhorn et al. 1995), but only one temperature-sensitive mutation in any histone gene has been reported (Smith et al. 1996). We will use the plasmid shuffle method to isolate both temperature-sensitive and galactose⁻ alleles of *HHT2* and/or *HHF2* .

STRAIN

7-2 UCC1118 *MATα ade2Δ::hisG his3Δ200 leu2Δ0 lys2Δ0 met15Δ0 trp1Δ63 ura3Δ0 Δhhf2-hht2::MET15 Δhhf1-hht1::LEU2 [* pRM200U (*URA3 CEN ARS HHF2-HHT2*)]

PLASMID

pMP3 A centromere vector containing *TRP1* and *HHT2-HHF2*.

PROCEDURE

Day 1.

Streak strain 7-2 onto a YPD plate. Incubate at 30°C.

Day 3.

Pick a robust colony of strain 7-2 and use it to inoculate a 5-ml YPD culture. Incubate overnight at 30°C.

Day 4.

Determine the cell density of the 5-ml strain 7-2 culture using the hemocytometer (see Appendix G). Dilute cells to 5 x 10^6 cells/ml in 50 ml of YPD and grow the culture for two additional divisions at 30°C (about 4 hours).

Harvest the cells and transform them, following Techniques and Protocols #1, High-efficiency Transformation of Yeast, with 100 ng of the mutagenized pMP3 plasmid (a stock will be provided that was made via Techniques and Protocols #6, Hydroxylamine Mutagenesis of Plasmid DNA). Make certain to include a no DNA control. Resuspend the transformed cells in 2.4 ml of sterile distilled H_2O. Remove 0.4 ml and dilute this to 2.0 ml. Plate 0.2-ml aliquots onto HC-trp media, using ten plates for the concentrated cells and another ten plates for the diluted cells. For the no DNA control, plate a 0.2-ml aliquot of concentrated cells on one plate. Incubate all 21 plates at *room temperature* (23°C).

Day 8.

Select ten transformation plates that ideally have 100–300 colonies each and determine the total number of colonies that will be assayed. Replica-plate each of the ten plates onto three 5-FOA plates, one 5-FOA+3%Galactose plate, and one HC-trp plate. Incubate one 5-FOA replica plate from each transformation at room temperature; incubate another 5-FOA and the 5-FOA+3%Galactose plate at 30°C; incubate the last 5-FOA replica plate and the HC-trp plate at 37°C.

Day 10.

Score the replica plates at 37°C for temperature-sensitive candidates. Pick as many as 12 temperature-sensitive candidates from the 5-FOA plate incubated at room temperature and purify by streaking onto YPD at room temperature (use a maximum of three YPD plates).

Day 12.

(i) Score the replica plates at 30°C (from Day 8) for colonies that are galactose$^-$. Pick as many as 12 galactose$^-$ candidates from the 5-FOA plate and purify by streaking on YPD at 30°C (use a maximum of three YPD plates). Note any galactose$^-$ colonies that might also be temperature-sensitive.

(ii) Retest the purified temperature-sensitive candidates (from Day 10) on YPD. Replica-plate each to two YPD plates, incubating one at room temperature and the other at 37°C .

Day 14.

(i) Retest the purified galactose$^-$ candidates (from Day 12). Replica each plate to one YPD and one YEPGal plate, incubating both at 30°C .

(ii) Score the re-tested temperature-sensitive candidates. Determine the frequency of temperature-sensitive plasmid transformants.

Day 16.

Score the re-tested galactose$^-$ candidates and determine the frequency of this class of transformants.

MATERIALS

Day 1. 1 YPD Plate

Day 3. 1 Culture tube containing 5 ml of YPD

Day 4. Materials for Techniques and Protocols #1, High-efficiency Transformation of Yeast
Mutagenized pMP3 DNA
50 ml of YPD in a flask
21 HC-trp plates

Day 8. 30 5-FOA plates
10 HC-trp plates
10 5-FOA+3% Galactose plates
Sterile velveteen pads

Day 10. 3 YPD plates

Day 12. 9 YPD plates
Sterile velveteen pads

Day 14. 3 YPD plates
3 YEPGal plates
Sterile velveteen pads

REFERENCES

Baudin, A., O. Ozier-Kalogeropoulous, A. Deanouel, F. LaCroute, and C. Cullin. 1993. A simple and efficient method for direct gene deletion in *Saccharomyces cerevisiae*. *Nucleic Acids Res.* **21:** 3329–3330.

Boeke, J.D., J. Trueheart, G. Natsoulis, and G.R. Fink. 1987. 5-Fluoro-orotic acid as a selective agent in yeast molecular genetics. *Methods Enymol.* **154:** 164–175.

Brachmann, C.B., A. Davies, G.J. Cost, E. Caputo, J. Li, P. Hieter, and J.D. Boeke. 1997. Designer deletion strains derived from *Saccharomyces cerevisiae* S288C: A useful set of strains and plasmids for PCR-mediated gene disruption and other applications. *Yeast* (in press).

Durrin, L.K., R.K. Mann, P.S. Kayne, and M. Grunstein. 1991. Yeast histone H4 N-terminal sequence is required for promoter activation in vivo. *Cell* **65:** 1023–1031.

Hirschhorn, J.N., A.L. Bortvin, S.L. Ricupero-Hovasse, and F. Winston. A new class of histone H2A mutations in Saccharomyces cerevisiae causes specific transcriptional defects in vivo. *Mol. Cell. Biol.* **15:** 1999–2009.

Lorenz, M.C., R.S. Muir, E. Lim, J. McElver, S.C. Weber, and J. Heitman. 1995. Gene disruption with PCR products in *Saccharomyces cerevisiae. Gene* **158:** 113–117.

Rose, M.D. 1987. Isolation of genes by complementation in yeast. *Methods Enzymol.* **152:** 481–504.

Rothstein, R.J. 1983. One-step gene disruption in yeast. *Methods Enzymol.* **101:** 202–211.

Scherer, S. and R.W. Davis. 1979. Replacement of chromosome segments with altered DNA sequences constructed in vitro. *Proc. Natl. Acad. Sci.* **76:** 4951–4955.

Smith, M.M., P. Yang, M.S. Santisteban, P.W. Boone, A.T. Goldstein, and P.C. Megee. 1996. A novel histone H4 mutant defective in nuclear division and mitotic chromosome transmission. *Mol. Cell. Biol.* **16:** 1017–1026.

Experiment VIII

Isolation of *ras2* Suppressors

RAS1 and *RAS2* specify very similar G-proteins that stimulate adenylyl cyclase in yeast. cAMP is essential for growth and for entry into the mitotic cell cycle. The components of the cAMP signal transduction pathway and their biochemical roles are well understood (see Broach 1991), based to a large extent on genetic analysis of the question of cell growth. In this experiment, we follow the approach of Cannon et al. (1986) to isolate suppressors of a *ras2* null mutation.

Under most circumstances, *ras1* or *ras2* single mutants are viable and healthy; one functional RAS protein is sufficient to stimulate adenylyl cyclase. However, *ras2* mutants are defective specifically in growth on nonfermentable carbon sources (such as acetate or glycerol). The reason is that *RAS1* expression is repressed in such growth media. Therefore, a *ras2* null mutation causes a conditional phenotype. We will exploit this phenotype to identify suppressors of *ras2*.

STRAINS

8-6	1784TRP	*MAT*α *RAS2 his4 ura3 leu2 can1*
8-7	AMP141	*MAT***a** *ras2-530::LEU2 his4 ura3 leu2 trp1 can1*
8-8	AMP142	*MAT*α *ras2-530::LEU2 his4 ura3 leu2 lys2 can1*

PROCEDURE

Day 1.

Instructors have streaked out your *ras2* mutant (strain 8-7 for groups 1, 2, 3, 4; strain 8-8 for groups 5, 6, 7, 8) on a YPD plate.

Construct a patch master on YPD with 30 to 40 isolated colonies. Incubate at 30°C. Refrigerate one of the streak plates to have a future source of the parent strain. Obtain a streak of the opposite strain (strain 8-8 for groups 1, 2, 3, 4; strain 8-7 for groups 5, 6, 7, 8) from a neighboring lab group and refrigerate for future use.

Day 2.

Replica-plate the YPD master plate to two YPAc plates. Replica-plate lightly to minimize background growth. Give the instructors one of the replicas for UV treatment (75 μJ setting in a Stratalinker). Incubate both plates at 30°C.

Day 6.

Papillae (small colonies growing out of the patches) should be visible. Pick one papilla from each patch, choosing either the UV-treated plate or the non-UV-treated plate in each case, and transfer to a fresh YPD master plate. Pick carefully to avoid contamination with the surrounding cells. (Normally, one would purify the papillae by streaking for single colonies, but we omit that step to save time.) Number each patch and record which plate the papilla came from. Include patches of the parent strain and of the *RAS2* control strain 8-6 on the master plate, and incubate it at 30°C.

Day 7.

Spread one lawn of the *ras2* strain of opposite mating type (strain 8-8 for groups 1, 2, 3, 4; strain 8-7 for groups 5, 6, 7, 8) on a YPD plate. Replica-plate the YPD master to two YPAc plates, and to two YPD plates and the *ras2* lawn. Incubate one YPAc/YPD pair at 30°C and the other at 37°C. Incubate these mating plates at 30°C.

Day 8.

Score growth on the 30°C and 37°C plates. Save these plates for use on Day 12. Choose 12 revertants that you want to analyze in detail. Streak out the crosses of these 12 revertants to the *ras2* tester strain on SC – lys – trp. Also streak out the cross of the *ras2* parent to the *ras2* tester strain. You should be able to streak out four crosses per plate. Incubate these plates at 30°C.

Day 10.

Pick two isolated colonies from each SC – lys – trp streak and frog to YPAc and YPD plates. Include a colony of the cross to the *ras2* parent. Incubate at 30°C.

Day 12.

Score growth on the YPD and YPAc plates to determine which suppressors are dominant and which are recessive. Go back to the YPD plate of Day 8 and prepare two stripe masters on YPD, each containing the *ras2* parent and 4 recessive suppressor strains. Incubate at 30°C.

Day 13.

Replica-plate each stripe master to two YPD plates. Replica-plate stripe masters from a friendly group of opposite mating type onto your replicas. Incubate at 30°C.

Day 14.

Replica-plate the crossed stripe masters to SC – lys – trp. Incubate at 30°C.

Day 15.

Pick cells from the stripe intersections and prepare a patch master on SC – lys – trp.

Day 16.

Replica-plate the Day 15 master plate to YPAc and to YPD. Incubate at 30°C.

Day 17.

Score the stripe-cross-masters from Day 16. What phenotype do you expect from suppressors that fail to complement? Are any of your mutants in the same complementation group as those from your neighbors? How many complementation groups are there?

MATERIALS

Day 1.	1 YPD plate Toothpicks
Day 2.	2 YPAc plates Sterile velveteen pads
Day 6.	1 YPD plate Toothpicks
Day 7.	4 YPD plates 2 YPAc plates Sterile velveteen pads Toothpicks
Day 8.	8 SC – lys – trp plates Toothpicks

Day 10. Toothpicks
1 YPAc plate
1 YPD plate
Sterile velveteen pads

Day 12. 2 YPD plates
Toothpicks

Day 13. 4 YPD plates
Sterile velveteen pads

Day 14. 4 SC – lys – trp plates
Sterile velveteen pads

Day 15. 1 SC – lys – trp plate
Toothpicks

Day 16. 1 YPAc plate
1 YPD plate
Sterile velveteen pads

Thanks to Rey Sia for designing the experiment.

REFERENCES

Broach, J.R. 1991. RAS genes in *Saccharomyces cerevisiae:* Signal transduction in search of a pathway. *Trends Genet.* **7:** 28–33.

Cannon, J.F., J.B. Gibbs, and K. Tatchell. 1986. Suppressors of the *ras2* mutation of *Saccharomyces cerevisiae. Genetics* **113:** 247–264.

EXPERIMENT IX

Manipulating Cell Types

Saccharomyces cerevisiae is capable of undergoing sexual reproduction. Haploid cells are able to mate with each other generating diploid cells. The diploid cells are then capable of undergoing meiosis to regenerate haploid cells. The ability to mate is determined by a cell's mating type. Cells of the same mating type are unable to mate, whereas cells of opposite mating type are able to mate. Mating type is determined by expression of alleles of the *MAT* locus. Cells of one mating type have the *MAT*α allele and are designated α cells, cells of the opposite mating type have the *MAT***a** allele and are designated **a** cells. Therefore, α cells are capable of mating with **a** cells, giving rise to diploids that contain both *MAT*α and *MAT***a** alleles. The resulting diploids are designated **a**/α.

The stability of mating type classifies strains into two groups, heterothallic and homothallic. Heterothallic strains have a stable mating type, whereas homothallic haploid strains are capable of changing mating type. Therefore, a colony derived from a homothallic cell will contain both **a** and α cells as well as diploid cells that arose from matings between **a** and α cells.

Spore-to-spore crosses between heterothallic and homothallic strains demonstrated that the distinction between homo- and heterothallism segregates as a single genetic locus designated *HO* (Winge and Roberts 1949; Takahashi et al. 1958). It was shown that a homothallic cell switches from one mating type to the other early in its cell lineage (Strathern and Herskowitz 1979).

A variety of genetic analyses demonstrated the presence of two other loci involved in homothallism (Takahashi 1958; Takano and Oshima 1967; Oshima and Takano 1972). These loci are now designated *HML* and *HMR* and are found with the *MAT* locus on chromosome III (Fig. 1). It has been shown that *HML* and *HMR* contain silent copies of

Figure 1 Diagrammatic representation of the silent loci HML and HMR in relation to the *MAT* locus on chromosome III.

the information expressed from the *MAT* locus (Hicks and Herskowitz 1977; Klar et al. 1979; Strathern et al. 1979). In general, *HML* has the α cell type information and *HMR* has the **a** cell type information. *HO* has been shown to encode a site-specific endonuclease that cuts at a site found specifically within the *MAT* locus. The double-strand break that occurs allows for recombination between the *MAT* locus and either *HML* or *HMR*. The recombination event, in general, occurs with the silent locus containing information for the opposite mating type from *MAT*. As this recombination event is not reciprocal, the information from the silent locus remains intact while the information at the *MAT* locus changes. These observations led to the cassette model for mating type switching (Fig. 2).

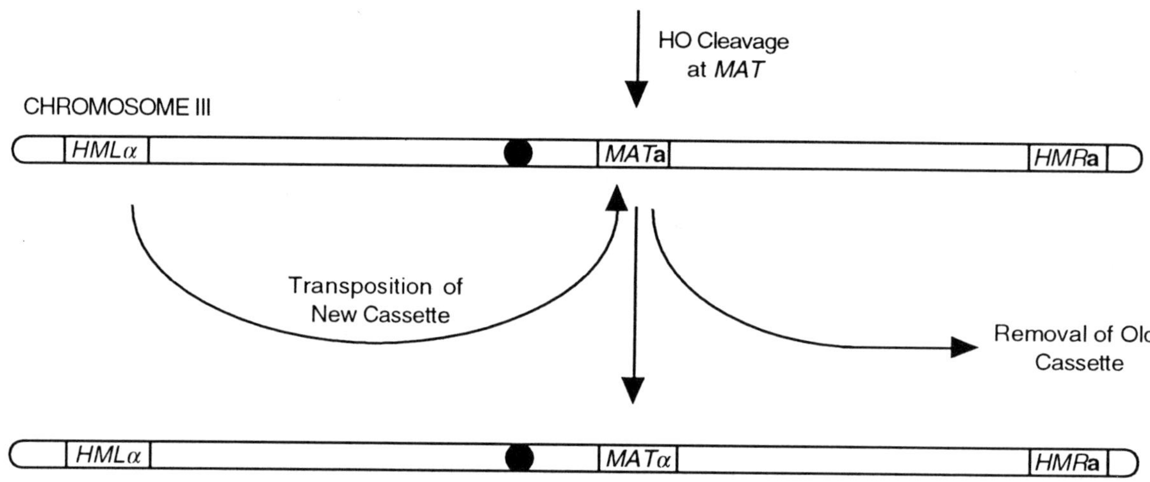

Figure 2 General model of mating type switching by the cassette model. In this case, α information replaces **a** at the *MAT* locus. (Adapted from Herskowitz and Oshima [1981], p. 194.)

In this experiment we will use: (a) *HO* to form diploids, and (b) a two-step gene replacement technique to switch the mating type of a strain.

EXPERIMENT IX(a)
Using *HO* To Generate Diploids

In this section, a haploid strain will be transformed with a plasmid containing *HO* to generate diploids. This procedure allows rapid purification of isogenic α and **a** haploid and α/**a** diploid cell types.

STRAINS

9-1	CKY8	*MATα ura3-52 leu2-3,112*
9-2	AAY1017	*MATα his1*
9-3	AAY1018	*MAT***a** *his1*

Note: Strains 9-2 and 9-3 are useful for testing mating types, but the genetic background is unknown. Therefore, these strains should never be used for genetic crosses other than mating-type testing.

PLASMID

pCY204 *HO* in YCp50. Selectable marker is *URA3* (Fig. 3).

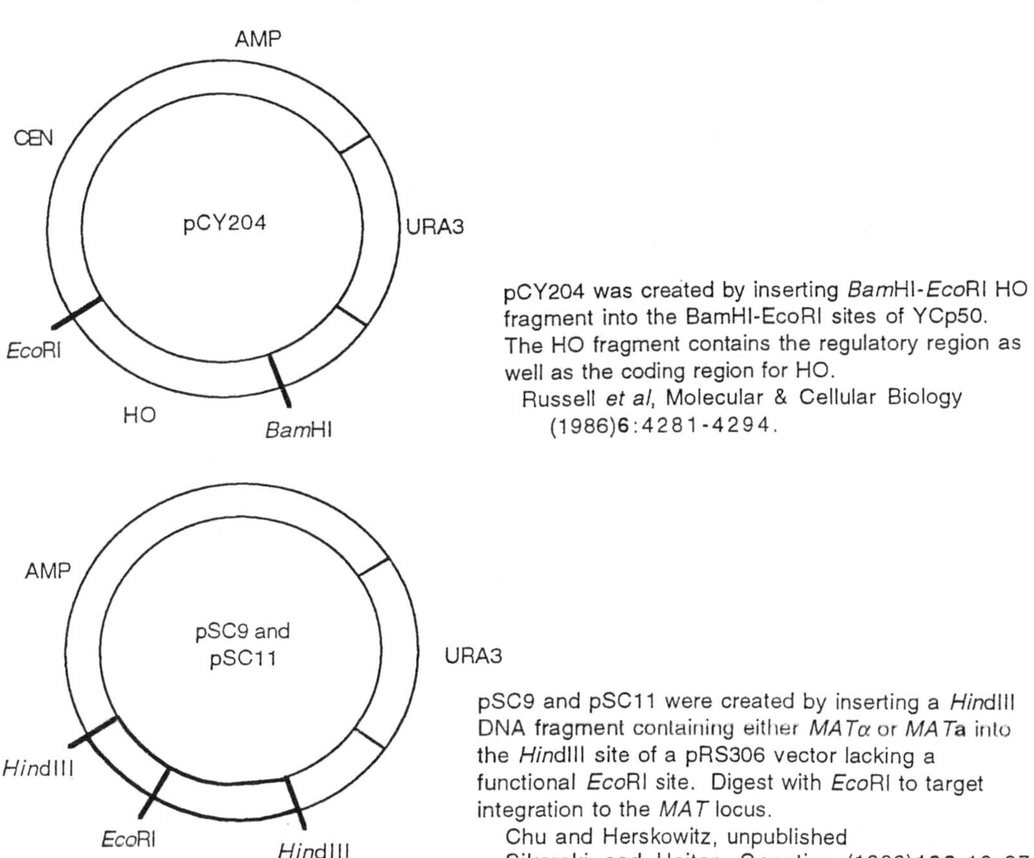

Figure 3 Plasmids for Experiment IX, Sections (a) and (b).

PROCEDURE

Day 1.

Subclone strain 9-1 onto a YPD plate and incubate at 30°C.

In the morning, inoculate 5 ml of YPD with a colony of strain 9-1 and incubate at 30°C. Store the YPD plate from above at 4°C for use on Day 11.

Day 2.

Transform strain 9-1 following Techniques & Protocols #1, High-efficiency Transformation of Yeast, with uncut plasmid pCY204. Plate transformants on SC - ura plates and incubate at 30°C. Remember to include a no-DNA control.

Day 5.

Streak 5 transformants for single colonies on SC - ura and grow at 30°C.

Day 7.

Inoculate 5 ml of YPD with a single colony from each individual transformant and grow at 30°C.

Day 8.

Determine cell density using a hemocytometer. Dilute cells appropriately in sterile distilled H_2O and plate 10^6 and 10^5 cells in 100-μl volumes on 5-FOA plates. This will select against the *URA3* gene so that cells that have lost pCY204 can grow. Store the remaining YPD cultures at 4°C for use on Day 11.

Day 11.

Make a patch master plate by patching nine 5-FOA^R colonies of each transformant onto a YPD plate (45 total). Also patch the parent strain 9-1. In addition, spot 3 μl of each of the five intermediate cultures from Day 7 onto the patch master plate. Incubate the YPD plate at 30°C. Inoculate two 5-ml YPD cultures with strain 9-2 and strain 9-3. Grow at 30°C with agitation.

Day 12.

To an SD plate, spread 100 µl of strain 9-2 and allow to dry. Repeat this process with another SD plate using strain 9-3. Replica-plate the patch master plate onto the SD plates with each lawn. Use a fresh velvet for each plate to prevent contamination between plates. Also replica-plate the patch master plate onto a sporulation plate, and onto an SC - ura plate.

Day 14.

Score mating ability and growth on the SC - ura plate.

Day 16.

Score sporulation ability.

MATERIALS

Day 1.
1 YPD plate
1 Culture tube containing 5 ml of YPD

Day 2.
Materials for Techniques and Protocols #1, High-efficiency Transformation of Yeast
Uncut plasmid pCY204
2 SC - ura plates

Day 5.
5 SC - ura plates

Day 7.
5 Culture tubes containing 5 ml of YPD

Day 8.
Sterile dH_2O
10 5-FOA plates

Day 11.
1 YPD plate
2 Culture tubes containing 5 ml of YPD

Day 12.
Sterile velveteen pads
2 SD plates
1 Sporulation plate
1 SC - ura plate

EXPERIMENT IX(b)

Mating Type Switching by Two-step Gene Replacement

Two-step gene replacement allows the replacement of one allele of a gene with another allele (see also Experiment VII). This technique is generally used to replace a wild-type allele with a mutant allele that has no selectable phenotype. The first step of this procedure involves integrating a plasmid sequence containing a selectable marker and the gene of interest. This type of integration leads to a duplication of the gene of interest, with the two genes separated by the plasmid sequence (Fig. 4A). The second step is loss of the plasmid sequence by homologous recombination between the duplicated regions. Loss of the plasmid is most easily accomplished if the marker used to

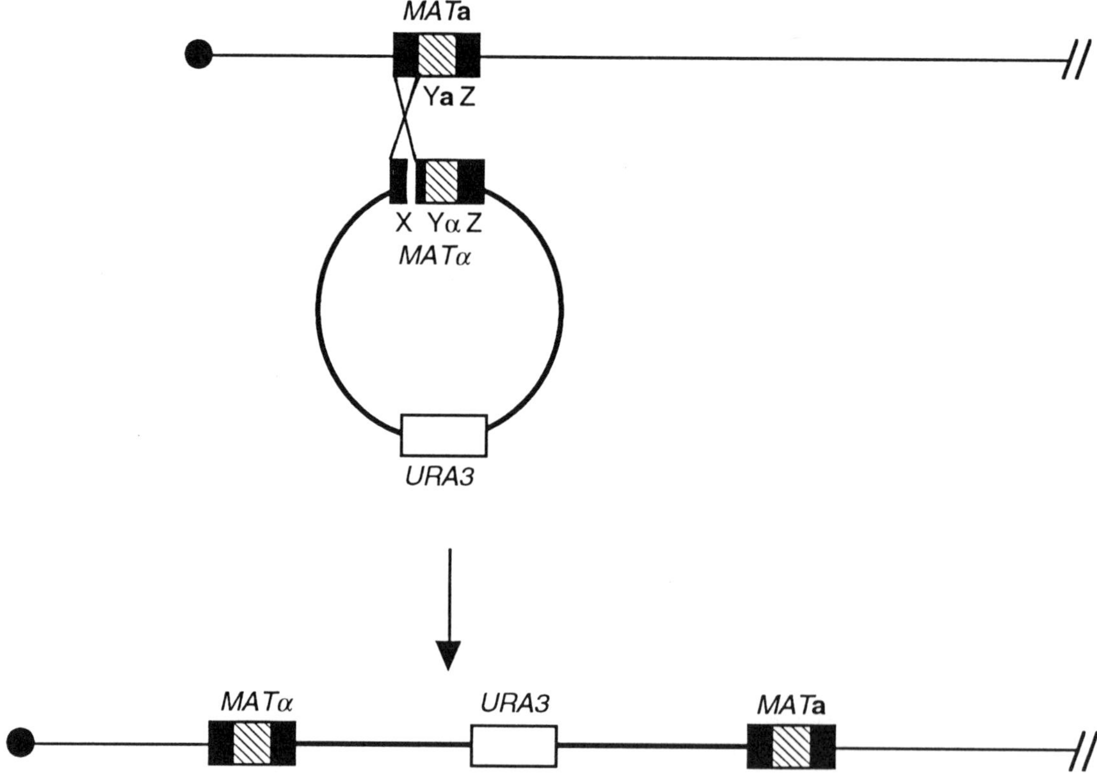

Figure 4A Two-step gene replacement.
Step 1: Integration of plasmid by homologous recombination at duplicated regions, either X or Z. Unlike one-step gene replacements, the entire plasmid is integrated. *Note:* The relative order of the two alleles is determined by the position of the crossover. In this figure, sequences determining mating-type occur 3′ of the crossover event.

select for integration can be selected against, for example, by selecting against *URA3* with 5-fluoro-orotic-acid (5-FOA). Depending on where the recombination event occurs, the remaining copy will carry one allele or the other (Fig. 4B). In this experiment, we will use two-step gene replacement to switch mating types (S. Chu and I. Herskowitz, unpubl.).

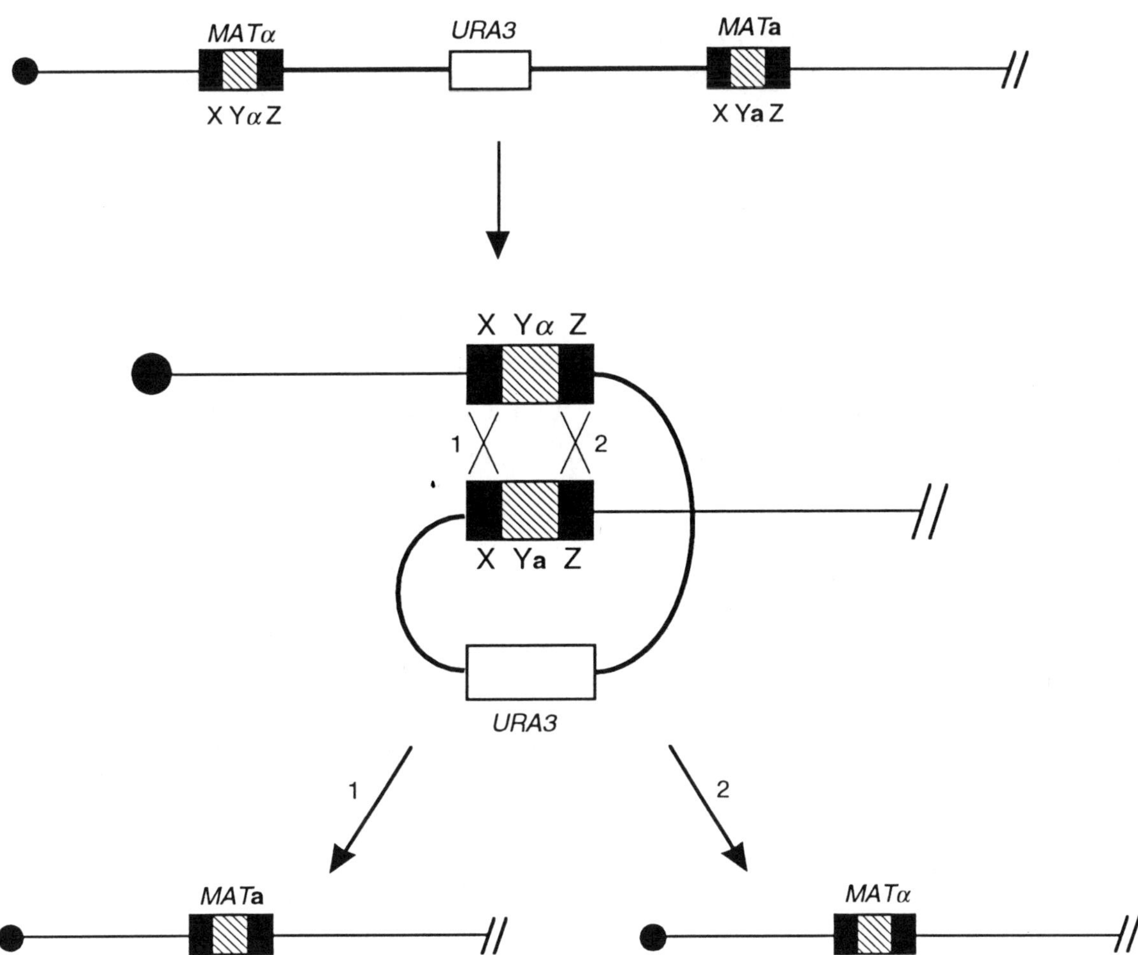

Figure 4B Two-step gene replacement (*continued*).
Step 2: Selection against the integration (in this case, selecting against *URA3* by plating on 5-FOA) allows for "looping out" by homologous recombination between the duplicated regions of the *MAT* locus (X or Z). Depending where recombination occurs (X or Z), the mating type will either be switched to *MAT*α or reverted to the parent allele (*MAT***a**).

STRAINS

9-4	YSC006	*MATα ura3 ade2-1 trp1-1 can1-100 leu2-3,112 his3-11,15[psi$^+$]GAL$^+$*
9-5	YSC005	*MAT**a** ura3 ade2-1 trp1-1 can1-100 leu2-3,112 his3-11,15[psi$^+$]GAL$^+$*

PLASMIDS

pSC9	*MATα* in pRS306 selectable marker is *URA3* (Fig. 4)
pSC11	*MAT**a*** in pRS306 selectable marker is *URA3* (Fig. 4)

PROCEDURE

Day 1.

Subclone strains 9-4 and 9-5 onto a YPD plate and incubate at 30°C. Groups 1, 2, 3, and 4 will use strain 9-4 and groups 5, 6, 7, and 8 will use strain 9-5.

In the morning, inoculate 5 ml of YPD with a single colony and incubate at 30°C. Store the YPD plate from above at 4°C for use on Day 11.

Day 2.

Transform strain 9-4 or 9-5 following Techniques & Protocols #1, High-efficency Transformation of Yeast. Obtain digested plasmid DNA from instructor for transformation. Strain 9-4 will be transformed with pSC11 and strain 9-5 with pSC9. Plate transformants on SC - ura plates and incubate at 30°C. Remember to include a no-DNA control.

Day 5.

Streak 5 transformants for single colonies on SC - ura and grow at 30°C.

Day 7.

Inoculate 5 ml of YPD with a single colony from each individual transformant and grow at 30°C.

Day 8.

Determine cell density using a hemocytometer (see Appendix G). Dilute cells appropriately in sterile distilled H_2O and plate 10^6 and 10^5 cells in 100-μl volumes on 5-

FOA plates. This will select against the *URA3* gene that should be inserted between the duplicated regions of the *MAT* locus. 5-FOA^R arises through ''looping out'' of the *URA3* gene by homologous recombination. Store the remaining YPD cultures at 4°C for use on Day 11.

Day 11.

Make patch master plate by patching nine 5-FOA^R colonies of each transformant onto a YPD plate (45 total). Also patch the parent strains 9-4 and 9-5. (Obtain one parent strain from the plate stored at 4°C on Day 1, and the other parent from a neighboring group.) In addition, spot 10 µl of each of the five intermediate cultures from Day 8 onto the patch master plate. Incubate the YPD plate at 30°C.

Day 12.

To an SD plate spread 100 µl of strain 9-2 (from Experiment IX[a], Day 11) and allow to dry. Repeat this process with another SD plate using strain 9-3. Replica-plate the patch master plate onto the SD plates with each lawn. Use a fresh velvet for each plate to prevent contamination between plates. Also replica-plate the patch master plate onto a sporulation plate, and onto an SC - ura plate.

Day 14.

Score mating ability and growth on the SC - ura plate.

Day 16.

Score sporulation ability.

MATERIALS

Day 1.	2 YPD plates
	1 Culture tube containing 5 ml of YPD
	Erlenmeyer flask containing 100 ml of YPD
Day 2.	Materials for Techniques and Protocols #1, High-efficency Transformation of Yeast
	Digested plasmid pSC9 and pSC11 DNA
	2 SC - ura plates
Day 5.	5 SC - ura plates

Day 7.	5 Culture tubes containing 5 ml of YPD
Day 8.	Sterile distilled H_2O
	10 5-FOA plates
Day 12.	1 YPD plate
	1 Sporulation plate
	1 SC - ura plate
Day 13.	Sterile velveteen pads
	2 SD plates

This experiment was developed and described by Dana Davis.

REFERENCES

Herskowitz, I. and Y. Oshima. 1981. Control of cell type in *Saccharomyces cerevisiae*: Mating type and mating-type interconversion. In *The molecular and cellular biology of the yeast* Saccharomyces: *Life cycle and inheritance* (ed. J.N. Strathern et al.), pp. 181–209. Cold Spring Harbor Laboratory, Cold Spring Harbor, New York.

Hicks, J.B. and I. Herskowitz. 1977. Interconversion of yeast mating types. II. Restoration of mating ability to sterile mutants in homothallic and heterothallic strains. *Genetics* **85:** 373.

Klar, A.J.S., S. Fogel, and D.N. Radin. 1979. Switching of a mating-type **a** mutant allele in budding yeast *Saccharomyces cerevisiae*. *Genetics* **92:** 759.

Oshima, Y. and I. Takano. 1972. Genetic controlling system for homothallism and a novel method for breeding triploid cells in *Saccharomyces*. *Fermentation Technology Today* 847.

Russell, D.W., R. Jensen, M.J. Zoller, J. Burke, B. Errede, M. Smith, and I. Herskowitz. 1986. Structure of the *Saccharomyces cerevisiae HO* gene and analysis of its upstream regulatory region. *Mol. Cell. Biol.* **6:** 4281–4294.

Sikorski, R.S. and P. Hieter. 1989. A system of shuttle vectors and yeast host strains designed for efficient manipulation of DNA in *Saccharomyces cerevisiae*. *Genetics* **122:** 19–27.

Strathern, J.N. and I. Herskowitz. 1979. Asymmetry and directionality in production of new cell types during clonal growth: The switching pattern of homothallic yeast. *Cell* **17:** 371–381.

Strathern, J.N., L.C. Blair, and I. Herskowitz. 1979. Healing of *mat* mutations and control of mating type interconversion by the mating type locus in *Saccharomyces cerevisiae*. *Proc. Natl. Acad. Sci.* **76:** 3425.

Takahashi, T. 1958. Complementary genes controlling homothallism in *Saccharomyces*. *Genetics* **43:** 705.

Takahashi, T., H. Saito, and Y. Ikeda. 1958. Heterothallic behavior of a homothallic strain in *Saccharomyces cerevisiae*. *Genetics* **43:** 249–260.

Takano, I. and Y. Oshima. 1967. An allele specific and a complementary determinant controlling homothallism in *Saccharomyces oviformis*. *Genetics* **57:** 875.

Winge, Ø. and C. Roberts. 1949. A gene for diploidization in yeast. *C. R. Trav. Lab. Carlsberg Ser. Physiol.* **24:** 341–346.

EXPERIMENT X

Isolation of Suppressors of Telomeric Silencing by Insertional Shuttle Mutagenesis

The eukaryotic genome is organized into regions distinct in their structure and function. Heterochromatin, which defines one such structural region, is condensed throughout the cell cycle, while its counterpart, euchromatin, is more diffuse in appearance during interphase. Chromosomal regions also differ functionally since the expression of a eukaryotic gene can be profoundly affected by its chromosomal position. This phenomenon, chromosomal position effect, has been observed in many eukaryotes (for review, see Lima-de-Faria 1983). When genetic rearrangements place euchromatic segments of the genome into or near heterochromatin, the expression of a translocated euchromatic gene is altered in a population of cells: some cells express the gene, while others do not. Thus, a mosaic or variegated phenotypic pattern is produced.

In *Saccharomyces cerevisiae*, three chromosomal domains have been shown to exert position effect: the rDNA locus, *HML, HMR*, and telomeres (Loo and Rine 1995; Bryk et al. 1997; Smith and Boeke 1997). Genes located near or within these domains can be transcriptionally silenced and exhibit phenotypic variegation. Telomeric silencing in *S. cerevisiae* appears to be caused by a repressive chromatin structure that initiates at the end of the chromosome and assembles inward along the DNA. A number of gene products involved in telomeric silencing have been identified, including histones H3 and H4. However, it is clear that the current collection of genes known to be important in telomeric chromatin structure and its assembly represents a fraction of the total required.

A SCREEN FOR SUPPRESSORS OF TELOMERIC SILENCING

In this experiment, a genetic screen will be conducted to identify additional components involved in telomeric silencing by isolating mutants that suppress telomeric silencing in a yeast strain that has been constructed with genetic markers located at two telomeric loci (Singer and Gottschling 1994). The *ADE2* gene, which is required for adenine biosynthesis, was placed adjacent to the telomere at the right arm of chromosome V (V-R), and *URA3*, a gene required for uracil biosynthesis, was located adjacent to the telomere at the left arm of chromosome VII (VII-L). Normally, colonies expressing *ADE2* are white, while those not expressing it (*ade2*) are red (Roman 1956). Due to the semi-stable nature of telomeric silencing of most genes, switching between silenced and transcriptionally active states may occur every few generations, giving rise to different phenotypic populations. In the case of strains with *ADE2* near a telomere, these different populations are seen as red and white sectors within a single colony (Gottschling et al. 1990). A *URA3* gene located at telomere VII-L also normally switches between transcriptional states. However, in this strain the telomeric *URA3* has been completely silenced by deleting its *trans*-activator, *PPR1* (Aparicio and Gottschling 1994). The cells are therefore unable to grow in the absence of uracil (Ura^-).

The screen for suppressors of telomeric silencing is carried out by mutagenizing this strain and identifying cells that no longer fully silence the two telomeric markers. That is, cells that are able to grow in the absence of uracil (Ura^+) and give rise to predominantly white colonies (Ade^+). The method of "insertional shuttle mutagenesis" will be used in this scheme because it provides a very powerful and effective way to quickly identify mutant genes in *S. cerevisiae* (Seifert et al. 1986; Burns et al. 1996). It takes advantage of yeast's highly efficient homologous recombination system and its fully sequenced, relatively small genome (~1.2×10^7 bp of haploid DNA).

Insertional shuttle mutagenesis is based on using a yeast::mTn3 library (Seifert et al. 1986). Basically, a large yeast genomic library containing 10^5 recombinants (~20 genome equivalents of yeast DNA) is constructed in an *E. coli* vector, with *Not*I restriction enzyme sites flanking the yeast genomic inserts. The library is then mutagenized by a Tn3 transposon derivative that contains a yeast selectable marker flanked by the Tn3 38 bp inverted repeat. DNA is then made from this mutagenized library and cleaved with *Not*I to release the mTn3-mutagenized yeast DNA from the vector. These linear DNA fragments are then transformed into a suitable strain selecting for the yeast marker to identify integrative recombinants.

In this study, the Tn3 derivative contains the yeast *LEU2* gene as the selectable marker, the bacterial *lacZ* gene, without an initiation codon, immediately adjacent to one of the 38 bp Tn3 repeats. One in six of Tn3 insertions within a gene can form a translational fusion with *lacZ*. Such fusions can be useful tools for additional analysis of the targeted yeast gene, and in this experiment a screen for such fusions will be carried out.

There are two ways of identifying the mutant gene, both of which take advantage of the inserted mTn3 transposon. One scenario involves rescue of the transposon and adjacent yeast DNA by a plasmid that is transferred from yeast to *E. coli* (Burns et al. 1996). In this experiment, yeast sequences flanking the mTn3 will be amplified using inverse-PCR (Ochman et al. 1990).

STRAINS

10-1 UCC3505 *MATα ura3-52 lys2-801 ade2-101 leu2-Δ1 his3-Δ200 Δppr1::LYS2 adh4::URA3-TEL-VIIL ADE2-TEL-VR*

PROCEDURE

Safety Note

The temperature of liquid nitrogen (N_2) is –185°C. Use with extreme caution. Always wear thermal gloves and face protection to prevent burns.

Day 1.

Streak strain 10-1 for single colonies onto a YPD plate. Incubate for 2 days at 30°C.

Day 3.

Pick a robust colony of strain 10-1 and use it to inoculate a 5-ml YPD culture. Incubate overnight at 30°C.

Day 4.

Determine the cell density of the 5-ml strain 10-1 culture using the hemocytometer (see Appendix G). Dilute cells to 5 x 10^6 cells/ml in an Erhlenmeyer flask containing 50 ml of YPD and grow culture for two additional divisions (about 4 hours) at 30°C.

Harvest the cells and transform them with 9 μg of *Not*I-cleaved mTn3-*lacZ-LEU2* genomic library DNA (the stock will be at 0.2 μg/μl), following Techniques and Protocols #1, High-efficiency Transformation of Yeast. Make certain to transform <u>nine times</u> with 1 μg of DNA per transformation and one transformation with <u>no</u> DNA, as a negative control. Plate each transformation onto three HC-leu plates, dividing the cells equally between the three plates. Incubate for 2 days at 30°C.

Day 6.

Determine the number of Leu^+ transformants. Replica-plate each HC-leu plate to an HC-leu-ura plate. Incubate for 3 days at 30°C.

Day 9.

Pick 20 independent Ura^+ colonies that are white (or at least not very pink) and streak for single colonies onto HC-ura plates. Be efficient and streak four colonies/plate, using a maximum of five plates. Incubate for 2 days at 30°C.

Day 11.

Determine the total number of independent transformants that re-tested to be Ura^+ and white (not very pink). Pick each one and patch it onto two YPD plates. Incubate for 1 day at 30°C. In addition, calculate the fraction of Leu^+ transformants that gave rise to Ura^+, white colonies. This is the frequency at which a relevant mutation occurred.

Day 12.

(i) Take one of the YPD plates and replica-plate it to a fresh YPD plate that has a sterile 3MW filter paper placed on top of the medium. Incubate overnight at 30°C.

(ii) Examine the other (master) plate and pick six colonies to inoculate independent 5-ml cultures of YPAD. Choose colonies that grew well on HC-ura media and did not show much pink or red color. These cultures will be used to isolate genomic DNA.

Day 13.

(i) Using forceps, carefully remove the 3MW filter paper from the YPD plate and immerse in liquid N_2 for 1 minute. **BE CAREFUL: Wear thermal gloves and eye protection to prevent burns.** Remove the filter, let it thaw completely, then repeat the liquid N_2 immersion one more time. After the filter thaws the second time, gently place it, with the cells facing upward, onto an X-gal plate. Incubate at 30°C and examine the plate every hour over the course of the day, noting any blue color development of the cells.

(ii) Isolate genomic DNA from the six cultures of cells using Techniques and Protocols #3D, Yeast Genomic DNA: Glass Bead Preparation, and resuspend the final DNA pellet in 100 μl of TE.

Day 14.

Digest 15 µl of each DNA (from Day 13) with 5–10 U of *Rsa*I in a total volume of 100 µl (make certain to add ~1 µg RNase A and the appropriate buffer to the digest) and incubate several hours at 37°C.

Inactivate the enzyme by heating the sample to 65°C for 20 minutes.

Remove 10 µl of the digested DNA and dilute to a final volume of 100 µl that is 1x T4 DNA ligase buffer (including ATP). Add ~100 U of T4 DNA ligase and incubate overnight at room temperature.

Day 15.

Amplify the DNA adjacent to the Tn3 transposon by PCR. To 5 µl of ligated DNA add:

4 µl of 25 mM $MgCl_2$
5 µl of 10x *Taq* Polymerase buffer
1 µl of 25 µM Primer #1
1 µl of 25 µM Primer #2
1 µl of 10 mM dNTPs
33 µl of distilled H_2O
1 µl of *Taq* Polymerase (2.5 U)

It is critical to use a "hot start" protocol for successful amplification of isolated genomic DNA. (Make certain that the sample is heated to 94°C for several minutes before adding *Taq* polymerase to the reaction.) Amplify with 35 cycles of 1 minute at 94°C, 1 minute at 65°C, and 2.5 minutes at 72°C. Conclude the final cycle with 7 minutes at 72°C.

Purify the amplified DNA away from protein, oligonucleotides, and dNTPs (a Wizard PCR Preps kit will be provided in the course), and submit for DNA sequencing.

Day 16.

Using a computer that is connected to the Internet, submit the DNA sequences for comparison to the yeast genome database and discover where your inserts landed!

```
  1 GGGGTCTGAC GCTCAGTGGA ACGAAAACTC ACGTTAAGGG GGATCCCGTC GTTTTACAAC GTCGTGACTG
    CCCCAGACTG CGAGTCACCT TGCTTTTGAG TGCAATTCCC CCTAGGGCAG CAAAATGTTG CAGCACTGAC
                   mTn3 sequencing primer (mTn3-SEQ)

 71 GGAAAACCCT GGCGTTACCC AACTTAATCG CCTTGCAGCA CATCCCCCTT TCGCCAGCTG GCGTAATAGC
    CCTTTTGGGA CCGCAATGGG TTGAATTAGC GGAACGTCGT GTAGGGGGAA AGCGGTCGAC CGCATTATCG
     Inverse PCR primer #1 (InPCR1)

141 GAAGAGGCCC GCACCGATCG CCCTTCCCAA CAGTTGCGCA GCCTGAATGG CGAATGGCGC TTTGCCTGGT
    CTTCTCCGGG CGTGGCTAGC GGGAAGGGTT GTCAACGCGT CGGACTTACC GCTTACCGCG AAACGGACCA

211 TTCCGGCACC AGAAGCGGTG CCGGAAAGCT GGCTGGAGTG CGATCTTCCT GAGGCCGATA CTGTCGTCGT
    AAGGCCGTGG TCTTCGCCAC GGCCTTTCGA CCGACCTCAC GCTAGAAGGA CTCCGGCTAT GACAGCAGCA

281 CCCCTCAAAC TGGCAGATGC ACGGTTACGA TGCGCCCATC TACACCAACG TAACCTATCC CATTACGGTC
    GGGGAGTTTG ACCGTCTACG TGCCAATGCT ACGCGGGTAG ATGTGGTTGC ATTGGATAGG GTAATGCCAG

351 AATCCGCCGT TTGTTCCCAC GGAGAATCCG ACGGGTTGTT ACTCGCTCAC ATTTAATGTT GATGAAAGCT
    TTAGGCGGCA AACAAGGGTG CCTCTTAGGC TGCCCAACAA TGAGCGAGTG TAAATTACAA CTACTTTCGA

421 GGCTACAGGA AGGCCAGACG CGAATTATTT TTGATGGCGT TAACTCGGCG TTTCATCTGT GGTGCAACGG
    CCGATGTCCT TCCGGTCTGC GCTTAATAAA AACTACCGCA ATTGAGCCGC AAAGTAGACA CCACGTTGCC

491 GCGCTGGGTC GGTTACGGCC AGGACAGTCG TTTGCCGTCT GAATTTGACC TGAGCGCATT TTTACGCGCC
    CGCGACCCAG CCAATGCCGG TCCTGTCAGC AAACGGCAGA CTTAAACTGG ACTCGCGTAA AAATGCGCGG

561 GGAGAAAACC GCCTCGCGGT GATGGTGCTG CGTTGGAGTG ACGGCAGTTA TCTGGAAGAT CAGGATATGT
    CCTCTTTTGG CGGAGCGCCA CTACCACGAC GCAACCTCAC TGCCGTCAAT AGACCTTCTA GTCCTATACA

631 GGCGGATGAG CGGCATTTTC CGTGACGTCT CGTTGCTGCA TAAACCGACT ACACAAATCA GCGATTTCCA
    CCGCCTACTC GCCGTAAAAG GCACTGCAGA GCAACGACGT ATTTGGCTGA TGTGTTTAGT CGCTAAAGGT
   Inverse PCR primer #2 (InPCR2)
                                              RsaI (738)
701 TGTTGCCACT CGCTTTAATG ATGATTTCAG CCGCGCTGTA CTGGAGGCTG AAGTTCAGAT GTGCGGCGAG
    ACAACGGTGA GCGAAATTAC TACTAAAGTC GGCGCGACAT GACCTCCGAC TTCAAGTCTA CACGCCGCTC
```

Figure 1 β-galactosidase end of mTn3-lacZ-LEU2 construct.

MATERIALS

Day 1. 1 YPD plate

Day 3. 1 Culture tube containing 5 ml of YPD

Day 4. Erlenmeyer flask containing 50 ml of YPD

Materials for Techniques and Protocols #1, High-efficiency Transformation of Yeast

0.5 mg of carrier DNA

9 μg of *Not*I-cleaved mTn3-*lacZ-LEU2* genomic library DNA
30 HC-leu plates

Day 6. 30 HC-leu-ura plates
30 Sterile velveteen pads

Day 9. 5 HC-ura plates

Day 11. 2 YPD plates

Day 12. 1 YPD plate
1 Sterile 3MW filter paper (8.2 cm in diameter)
6 Culture tubes containing 5 ml of YPAD

Day 13. Liquid N_2
1 X-Gal plate
Materials for Techniques and Protocols #3D, Yeast Genomic DNA: Glass Bead Preparation

Day 14. *Rsa*I
60 μl of 10x Restriction enzyme buffer
5 μl of RNase A (5 mg/ml)
60 μl of 10x T4 DNA Ligase buffer + ATP
T4 DNA Ligase

Day 15. 50 μl of 25 mM $MgCl_2$
50 μl of 10x *Taq* Polymerase buffer
10 μl of 25 μM Primer #1
10 μl of 25 μM Primer #2
10 μl of 10 mM dNTPs
Distilled H_2O
Taq Polymerase
Wizard™ PCR Preps DNA Purification System (Promega Corp. A 7170)
ABI Dye Terminator Sequencing Kit with mTn3 sequencing primer (Applied Biosystems Inc.)

Day 16. Networked computer

REFERENCES

Aparicio, O.M. and D.E. Gottschling. 1994. Overcoming telomeric silencing: A trans-activator competes to establish gene expression in a cell cycle-dependent way. *Genes Dev.* **8:** 1133–1146.

Bryk, M., M. Banerjee, M. Murphy, K.E. Knudsen, D.J. Garfinkel, and M.J. Curcio. 1997. Transcriptional silencing of TY1 elements in the RDN1 locus of yeast. *Genes Dev.* **11:** 255–269.

Burns, N., P. Ross-Macdonald, G.S. Roeder, and M. Snyder. 1996. *Methods in Molecular Genetics* **5:** 298–308.

Gottschling, D.E., O. M. Aparicio, B.L. Billington, and V.A. Zakian. 1990. Position effect at *S. cerevisiae* telomeres: Reversible repression of Pol II transcription. *Cell* **63**: 751–762.

Lima-de-Faria, A. 1983. *Molecular evolution and organization of the chromosome*, p. 507. Elsevier Science Publishers B.V., Amsterdam, The Netherlands.

Loo, S. and J. Rine. 1995. Silencing and heritable domains of gene expression. *Ann. Rev. Cell Dev. Biol.* **11**: 519–548.

Ochman, H., M.M. Medhora, D. Garza, and D.L. Hartl. 1990. *PCR Protocols:A Guide to Methods and Applications*, p. 219. Academic Press.

Roman, H. 1956. Studies of gene mutation in *Saccharomyces. Cold Spring Harbor Symp. Quant. Biol.* **21:** 175–185.

Singer, M.S. and D.E. Gottschling. 1994. TLC1: Template RNA component of *Saccharomyces cerevisiae* telomerase. *Science* **266:** 404–409.

Seifert, H.S., E.Y. Chen, M. So, and F. Heffron. 1986. Shuttle mutagenesis: A method of transposon mutagenesis for *Saccharomyces cerevisiae. Proc. Natl. Acad. Sci.* **83:** 735–739.

Smith, J.S. and J.D. Boeke. 1997. An unusual form of transcriptional silencing in yeast ribosomal DNA. *Genes Dev.* **11:** 241–254.

EXPERIMENT XI

Two-hybrid Protein Interaction Method

Most biological processes are mediated by protein-protein interactions, and a wealth of biochemical assays have been developed to detect such interactions. The two-hybrid system is a yeast-based genetic assay that provides a simple and sensitive means to detect potential interactions between two proteins. It is based on the finding that certain eukaryotic transcription activators are modular. For example, Gal4p, a positive regulator of the *GAL* pathway, consists of a site-specific DNA-binding domain and an acidic transcription activation domain (Keegan et al. 1986). The DNA-binding domain binds to an upstream activating sequence (UAS) found in the promoters of GAL genes. The transcription activation domain interacts with other components of the transcription apparatus to initiate transcription. These two domains are separable and can function as independent units. Fields and Song (1989) used this feature of Gal4p to create a system in which two hybrid proteins are created: one protein is fused to the Gal4p DNA-binding domain, and another is fused to the Gal4p transcription activation domain. The binding and activation domains themselves do not interact with each other, and when expressed together in a yeast cell are unable to activate transcription at a *GAL* promoter. However, if the sequences fused to the binding and activation domains are able to interact with each other, then the binding and activation domains are brought together on the DNA and a functional transcriptional activator is reconstituted.

In the simplest case, the two-hybrid system is used to test for interactions between two known proteins; this is often referred to as a "directed" two-hybrid. A variation of the directed two-hybrid is to test the ability of mutant forms (typically deletions) of the proteins of interest to interact, potentially identifying domains required for the interaction. Another use of the two-hybrid system is to identify interacting proteins from a random genomic or cDNA library. In this case, the protein of interest is fused to the DNA-binding domain, and random genomic or cDNA fragments are fused to the activation domain. In the experiment described here, a variation of the original yeast two-hybrid system developed by Elledge (Harper et al. 1993) will be used to examine the interaction between two proteins, Cin2p and Cin4p. *CIN2* and *CIN4* are involved in yeast microtubule function (Hoyt et al. 1990; Stearns et al. 1990), and have been shown to interact in the two-hybrid system (Feierbach and Stearns, unpubl.).

The activity of the hybrid transcriptional activator is assayed using constructs that have a GAL promoter driving one of a variety of reporter genes. In the version used in this experiment, the two reporter genes are *HIS3*, a yeast gene involved in histidine biosynthesis, and *lacZ*, the bacterial gene for β-galactosidase. The *HIS3* reporter allows for selection of His^+ cells on medium lacking histidine. The *lacZ* reporter allows for visual screening of cells expressing b-galactosidase, as visualized with the chromogenic substrate X-gal. The promoters driving these reporters differ in the number and affinity of upstream binding sites for the DNA-binding hybrid and in the position of these sites relative to the transcription start-point. These differences affect the strength of the protein interaction the reporters can detect; activation of the *lacZ* reporter is the more rigorous test for function.

Several requirements must be fulfilled for the two-hybrid method to succeed. First, both hybrid proteins must be expressed. In some versions of the two-hybrid vectors, an epitope tag is included in the construct, allowing expression to be verified by Western blotting. Second, neither hybrid protein should be able to activate transcription by itself. When identifying interactors from a library, there are several possibilities for false positives, not all of which are understood mechanistically. To help to identify such false positives, plasmids that are recovered from the library are subjected to additional tests: first, for self-activation of the reporter genes, then for specificity of interaction with protein of interest, as compared to several standard test proteins (p53 and lamin are typically used). As with all genetic methods, interactions detected in the two-hybrid system should be confirmed by biological and/or biochemical experiments.

STRAINS

11-1	Y190	*MAT***a** *gal4 gal80 his3-Δ200 trp1-901 ade2-101 leu2-3,112 lys2:GAL-HIS3:LYS2 ura3-52:GAL-lacZ:URA3* Cyh^r
11-2	Y187	*MATα gal4 gal80 his3-Δ200 trp1-901 ade2-101 ura3-52 leu2-3,112 ura3-52:GAL-lacZ:URA3*
11-3	TSY801	Y190 [pTS434]
11-4	TSY802	Y187 [pTS721]
11-5	TSY803	Y190 [pAS1]

11-6	TSY804	Y187 [pACTII]
11-7	TSY805	Y190 [pTS428]
11-8	TSY806	Y190 [pTS430]

PLASMIDS

pAS1	GAL4 DNA-binding domain vector (TRP1 is selectable marker)
pACTII	GAL4 activation domain vector (LEU2 is selectable marker)
pTS434	CIN4 in pAS1 (DNA-binding domain fusion)
pTS721	CIN2 in pACTII (GAL4 activation domain fusion)
pTS428	pAS1-p53
pTS430	pAS1-lamin

PROCEDURE

Safety Note

The temperature of liquid nitrogen (N_2) is –185°C. Use with extreme caution. Always wear thermal gloves and a face protection to prevent burns.

Day 1.

Patch out all strains onto SC-leu and SC-trp, depending on the selectable marker of the plasmid in each strain. Incubate at 30°C.

Day 3.

In order to express the various DNA-binding domain fusions and activation domain fusions in the same cell, strains carrying each plasmid will be mated pairwise, and the resulting diploids assayed for reporter gene activation.

Genetic crosses. Make pairwise crosses between strains in the combinations in the table below. Use sterile toothpicks to mix some cells of each of the two strains in a spot on a YPD plate. Incubate overnight at 30°C.

Strain 1	Strain 2	Selective medium
11-3	11-4	SD+Ade
11-3	11-6	SD+Ade
11-4	11-5	SD+Ade
11-4	11-7	SD+Ade
11-4	11-8	SD+Ade

Day 4.

Streak each cross onto SD+ ade plates to select for diploids. Incubate overnight at 30°C for 2 days.

Day 6.

(i) Streak each diploid onto SD+ ade+25 mM 3-aminotriazole (3-AT) plates to test for activation of the *HIS3* gene. Incubate overnight at 30°C. Note that 3-aminotriazole is a competitive inhibitor of the His3p enzyme, and is necessary to make the His^+ selection more stringent.

(ii) Patch each diploid onto SD+ade plates for X-gal filter lift assay. Incubate overnight at 30°C.

Day 7.

(i) Check for growth on plates containing 3-AT. Growth demonstrates activation of the *HIS3* gene. Depending on the diploid tested, this activation can indicate an interaction or self-activation.

(ii) X-gal Filter-Lift Assay:

1. Using a pencil or ballpoint pen, label 45-µm circular nitrocellulose filters.
2. Lay a filter onto the plate with the patched yeast strains and allow it to wet completely. Avoid getting bubbles between the filter and plate.

3. Lift the filter off the plate carefully to avoid smearing the patches and place the filter in liquid N_2 to permeabilize the cells (5–10 seconds is sufficient).
4. Carefully remove the filter from the liquid N_2 (the frozen filters are very brittle) and place cell-side-up in a petri dish containing 3MM chromatography paper soaked in 3 ml of Z buffer containing 1 mg/ml X-Gal.
5. Incubate for several minutes (or up to several hours) at 30°C for development of blue color indicating β-galactosidase activity.

Day 8.

Check the 3-AT plates again for growth.

MATERIALS

Day 1. 1 SC-leu plate
1 SC-trp plate

Day 3. 2 YPD plates

Day 4. 3 SD+ade plates

Day 6. 3 SD+ade+25 mM 3-AT plates
1 SD+ade plate

Day 7. 1 Nitrocellulose filter (45 μm; Schleicher & Schuell BA85)
3 ml of Z buffer with 1 mg/ml X-gal
Liquid N_2

REFERENCES

Harper, J.W., G.R. Adami, N. Wei, K. Keyomarski, and S.J. Elledge, 1993. The p21 Cdk-interacting protein Cip1 is a potent inhibitor of G1 cyclin-dependent kinases. *Cell* **75:** 805–816.

Fields, S. and O. Song. 1989. A novel genetic system to detect protein-protein interactions. *Nature* **340:** 245–246.

Hoyt, M.A., T. Stearns, and D. Botstein. 1990. Chromosome instability mutants of *Saccharomyces cerevisiae* that are defective in microtubule-mediated processes. *Mol. Cell. Biol.* **10:** 223–234.

Hoyt, M.A., T. Stearns, and D. Botstein. 1990. Chromosome instability mutants of *Saccharomyces cerevisiae* that arc dcfective in microtubule-mediated processes. *Mol. Cell. Biol.* **10:** 223–234.

Keegan, L., G. Gill, and M. Ptashne. 1986. Separation of DNA binding from the transcription-activating function of a eukaryotic regulatory protein. *Science* **231:** 699–704.

Stearns, T., M.A. Hoyt, and D. Botstein. 1990. Yeast mutants supersensitive to antimicrotubule drugs define three genes that affect microtubule function. *Genetics* **124:** 251–262.

TECHNIQUES & PROTOCOLS #1

High-efficiency Transformation of Yeast

(Adapted from Gietz and Schiestl [1995] *Methods in Molecular and Cellular Biology* **5[5]**: 255–269.)

1. Inoculate 5 ml of liquid YPAD or 10 ml of SC and incubate with shaking overnight at 30°C.

2. Count overnight culture and inoculate 50 ml of YPAD to a cell density of 5 x 10^6/ml culture.

3. Incubate the culture at 30°C on a shaker at 200 rpm until it is at 2 x 10^7 cells/ml. This typically takes 3–5 hours. This culture will give sufficient cells for 10 transformations.

   Notes:
   i. It is important to allow the cells to complete at least 2 divisions.
   ii. Transformation efficiency remains constant for 3 to 4 cell divisions.

4. Harvest the culture in a sterile 50-ml centrifuge tube at 3000*g* (2500 rpm) for 5 minutes.

5. Pour off the medium, resuspend the cells in 25 ml of sterile H_2O, and centrifuge again.

6. Pour off the H_2O, resuspend the cells in 1.0 ml of 100 mM lithium acetate (LiAc), and transfer the suspension to a sterile 1.5-ml microfuge tube.

7. Pellet the cells at top speed for 5 seconds and remove the LiAc with a micropipette.

8. Resuspend the cells to a final volume of 500 μl (2 x 10^9 cells/ml), which is about 400 μl of 100 mM LiAc.

   Note: If the cell titer of the culture is greater than 2 x 10^7 cells/m, the volume of the LiAc should be increased to maintain the titer of this suspension at 2 x 10^9 cells/ml. If the titer of the culture is less than 2 x 10^7 cells/ml, then decrease the amount of LiAc.

9. Boil a 1.0-ml sample of single-stranded carrier DNA for 5 minutes and quickly chill in ice water.

Note: It is not necessary or desirable to boil the carrier DNA every time. Keep a small aliquot in your own freezer box and boil after 3–4 freeze-thaws. But keep on ice when out.

10. Vortex the cell suspension and pipette 50-μl samples into labeled microfuge tubes. Pellet the cells and remove the LiAc with a micropipette.

11. The basic "transformation mix" consists of the following ingredients; carefully add them *in the order listed:*

    240 μl of PEG (50% w/v)
    36 μl of 1.0 M LiAc
    25 μl of Single-stranded carrier DNA (2.0 mg/ml)
    50 μl of H_2O and Plasmid DNA (0.1–10 μg)

Note: The order is important here! The PEG, which shields the cells from the detrimental effects of the high concentration of LiAc, should go in first.

12. Vortex each tube vigorously until the cell pellet has been completely mixed. This usually takes about 1 minute.

13. Incubate for 30 minutes at 30°C.

14. Heat shock for 20–25 minutes in a water bath at 42°C.

    Note: The optimum time can vary for different yeast strains. Please test this if you need high efficiency from your transformations.

15. Microfuge at 6000–8000 rpm for 15 seconds and remove the transformation mix with a micropipette.

16. Pipette 0.2-1.0 ml of sterile H_2O into each tube and resuspend the pellet by pipetting it up and down gently.

    Note: Be as gentle as possible at this step if high efficiency is important.

17. Plate from 200-μl aliquots of the transformation mix onto selective plates.

MATERIALS AND SOLUTIONS

Polyethylene glycol (PEG; 50% w/v) (MW 3350; Sigma P 3640)

Make up to 50% (w/v) with H_2O and filter-sterilize with a 0.45-µm filter unit (Nalgene). Alternatively, the PEG solution can be autoclaved, but care must be taken to ensure that the PEG solution is at the proper concentration. In addition, it is important to store the PEG in a tightly-capped container to prevent evaporation of H_2O and a subsequent increase in PEG concentration. Small variations above or below the PEG concentration optimum in the transformation reaction, which is 33% (w/v), can reduce the production of transformants.

Single-stranded Carrier DNA (2 mg/ml)

High-molecular-weight DNA (Deoxyribonucleic acid Sodium Salt Type III from Salmon Testes; Sigma D 1626)

TE buffer (pH 8.0)

10 mM Tris-HCl (ph 8.0)

1.0 mM EDTA

1. Weigh out 200 mg of the DNA into 100 ml of TE buffer. Disperse the DNA into solution by drawing it up and down repeatedly in a 10-ml pipette. Mix vigorously on a magnetic stirrer for 2–3 hours or until fully dissolved. Alternatively, leave the covered solution mixing at this stage overnight in a cold room.
2. Aliquot the DNA (100 µl is typically convenient), and store at –20° C.
3. Prior to use, an aliquot should be placed in a boiling water bath for at least 5 minutes and quickly cooled in an ice water slurry.

Tips:

i. Carrier DNA can be frozen after boiling and used 3 or 4 times. If transformation efficiencies begin to decrease with a batch of boiled carrier DNA, it should be boiled again or a new aliquot used.

ii. The lower concentration of carrier DNA (2 mg/ml) in this protocol eases handling and gives more reproducible results.

iii. In previous protocol versions, a phenol : chloroform extraction was used to ensure maximal transformation efficiencies. This extraction may not be necessary if the DNA is of high enough quality. Test your carrier DNA to determine if extraction is necessary.

1.0 M Lithium acetate stock solution (LiAc)

Prepare as a 1.0 M stock in distilled deionized H_2O; filter-sterilize. There is no need to titrate this solution, but the final pH should be between 8.4 –8.9.

The latest version of this method can be found at:
http://www.umanitoba.ca/faculties/medicine/human_genetics/gietz/method.html

REFERENCE

R.D. Gietz and R.H. Schiestl. 1995. Transforming yeast with DNA. *Methods Mol. Cell. Biol.* **5(5):** 255–269.

TECHNIQUES & PROTOCOLS #2

"Lazy Bones" Plasmid Transformation of Yeast Colonies

(Modified from Elble [1992] *BioTechniques* **13[1]:** 18–20.)

1. Pick a colony (2–3 mm in diameter) from a plate with a toothpick and transfer cells to a sterile 1.5-ml microfuge tube.
2. Add 10 μl of carrier DNA (100 μg) plus transforming plasmid DNA (in 10 μl) and vortex well. (Carrier DNA does not need to be added if the transforming DNA has come from mini-prep DNA that has not been RNased.)
3. Add 0.5 ml of PLATE solution and vortex.
4. Incubate overnight–4 days at room temperature on the bench top.
5. Heat shock for 15 minutes at 42°C.
6. Pellet cells for 10 seconds at 8–10 krpm in a microfuge. Carefully remove liquid and gently resuspend cells in 200 μl of sterile distilled H_2O by pipetting up and down. Spread mixture directly onto selective plates.

MATERIALS AND SOLUTIONS

PLATE solution

- 40% Polyethylene glycol (PEG) (MW 3350; Sigma P 3640)
- 0.1 M Lithium acetate (LiAc)
- 10 mM Tris-HCl (pH 7.5)
- 1 mM EDTA

Carrier DNA (10 μg/μl)

Tip:

Cells from a fresh plate transform very efficiently; old cells that have been on plates several months can also be transformed, though less efficiently.

REFERENCE

Elble, R. 1992. A simple and efficient procedure for transformation of yeasts. *BioTechniques* **13 (1):** 18–20.

TECHNIQUES & PROTOCOLS #3

Yeast DNA Isolations

A. YEAST DNA MINIPREP (40 ml)

1. Grow cells at 30°C to saturation (overnight) in 40 ml of YPD in a 125-ml flask.

2. Centrifuge the cells in a clinical centrifuge or a Sorvall SS-34 rotor at 5000 rpm for 5 minutes in a screw-capped centrifuge tube. Discard the supernatant.

3. Resuspend the cells in 3 ml of 0.9 M sorbitol, 0.1 M Na_2 EDTA (pH 7.5).

4. Add 0.1 ml of a 2.5 mg/ml solution of Zymolyase 100T and incubate for 1 hour at 37°C.

5. Centrifuge the cells in a clinical centrifuge or Sorvall SS-34 rotor for 5 minutes at 5000 rpm. Discard the supernatant.

6. Resuspend the cell pellet in 5 ml of 50 mM Tris-Cl (pH 7.4), 20 mM Na_2 EDTA.

7. Add 0.5 ml of 10% SDS and mix.

8. Incubate for 30 minutes at 65°C.

9. Add 1.5 ml of 5 M potassium acetate and store on ice for 1 hour.

10. Centrifuge in a Sorvall SS-34 rotor at 10,000 rpm for 10 minutes.

11. Transfer the supernatant to a fresh plastic centrifuge tube and add two volumes of 95% ethanol at room temperature. Mix and centrifuge at 5000–6000 rpm for 15 minutes at room temperature.

12. Discard the supernatant. Dry the pellet and then resuspend in 3 ml of TE (pH 7.4). This may take several hours.

13. Centrifuge in a Sorvall SS-34 rotor at 10,000 rpm for 15 minutes and transfer the supernatant to a new tube. Discard the pellet.

14. Add 150 μl of a 1 mg/ml solution of RNase A and incubate for 30 minutes at 37°C.

15. Add one volume of 100% isopropanol and shake gently to mix. Remove the precipitate, which should now look like a loose ''cocoon'' of fibers. Do not centrifuge. Air dry.

16. Resuspend the precipitate in 0.5 ml of TE (pH 7.4). Store at 4°C. The final concentration of yeast DNA should be ~200 μg/ml. If the final solution is milky, reprecipitate the DNA with isopropanol or centrifuge in a Sorvall SS-34 rotor at 10,000 rpm for 15 minutes.

MATERIALS AND SOLUTIONS

125-ml flask containing 40 ml of YPD

Screw-capped centrifuge tubes

0.9 M Sorbitol, 0.1 M Na_2 EDTA (pH 7.5) (3 ml)

Zymolyase® 100T (120493-1, Seikagaku America Inc.) solution (0.1 ml)

  2.5 mg/ml in 0.9 M sorbitol, 0.1 M Na_2 EDTA (pH 7.5)

50 mM Tris-Cl (pH 7.4), 20 mM Na_2 EDTA (5 ml)

10% SDS

5 M Potassium acetate (1.5 ml)

95% Ethanol

TE (pH 7.4)

  10 mM Tris-Cl (pH 7.4)

  1 mM Na_2 EDTA

RNase A solution (150 μl)

  Dissolve at a concentration of 1 mg/ml in 50 mM potassium acetate (pH 5.5). Boil for 10 minutes. Store frozen at –20°C.

100% Isopropanol

B. YEAST DNA MINIPREP (5 ml)

1. Grow cells overnight at 30°C in 5 ml of YPD.

2. Collect the cells in a clinical centrifuge at 2000 rpm for 5 minutes. Discard the supernatant.

3. Resuspend the cells in 0.5 ml of 1 M sorbitol, 0.1 M Na_2 EDTA (pH 7.5); transfer to a 1.5-ml microfuge tube.

4. Add 0.02 ml of a 2.5 mg/ml solution of Zymolyase 100T and incubate for 1 hour at 37°C.

5. Centrifuge in a microfuge for 1 minute.

6. Discard the supernatant. Resuspend the cells in 0.5 ml of 50 mM Tris-Cl (pH 7.4), 20 mM Na_2 EDTA.

7. Add 0.05 ml of 10% SDS; mix well.

8. Incubate the mixture for 30 minutes at 65°C.

9. Add 0.2 ml of 5 M potassium acetate and place the microfuge tube in ice for 1 hour.

10. Centrifuge in microfuge for 5 minutes.

11. Transfer the supernatant to a fresh microfuge tube and add one volume of 100% isopropanol at room temperature. Mix and allow it to sit at room temperature for 5 minutes. Centrifuge VERY BRIEFLY (10 seconds) in a microfuge. Pour off the supernatant and air-dry the pellet.

12. Resuspend the pellet in 0.3 ml of TE (pH 7.4).

13. (Optional) Add 15 μl of a 1 mg/ml solution of RNase A and incubate for 30 minutes at 37°C.

14. Add 0.03 ml of 3 M sodium acetate and mix. Precipitate with 0.2 ml of 100% isopropanol. Centrifuge briefly again to collect the pellet of DNA.

15. Pour off the supernatant and air-dry. Resuspend the pellet of DNA in 0.1–0.3 ml of TE (pH 7.4).

16. Before using the DNA solution in a restriction digest, it may be necessary to centrifuge the final solution hard (15 minutes) in a microfuge to remove insoluble material that may inhibit digestion.

MATERIALS AND SOLUTIONS

YPD

1 M Sorbitol, 0.1 M Na_2 EDTA (pH 7.5) (0.5 ml)

Zymolyase® 100T (120493-1, Seikagaku America Inc.) solution (0.02 ml)

  2.5 mg/ml in 1 M sorbitol, 0.1 M Na_2 EDTA (pH 7.5)

50 mM Tris-Cl (pH 7.4), 20 mM Na_2 EDTA (0.5 ml)

10% SDS

5 M Potassium acetate (0.2 ml)

100% Isopropanol

TE (pH 7.4)

  10 mM Tris-Cl (pH 7.4)

  1 mM Na_2 EDTA

RNase A solution (15 μl) (optional)

  Dissolve at a concentration of 1 mg/ml in 50 mM potassium acetate (pH 5.5). Boil for 10 minutes. Store frozen at –20°C.

3 M Sodium acetate (0.03 ml)

C. A TEN-MINUTE DNA PREPARATION FROM YEAST

(From Hoffman and Winston *Gene* **57:** 267–272 [1987].)

Safety Notes

Phenol is highly corrosive and can cause severe burns. Gloves, protective clothing, and safety glasses should be worn when handling it. All manipulations should be carried out in a chemical hood. Any areas of skin that come in contact with phenol should be rinsed with a large volume of water or polyethylene glycol 400 and washed with soap and watcr; ethanol should not be used.

Chloroform is irritating to the skin, eyes, mucus membranes, and upper respiratory tract. It should only be sued in a chemical hood. Gloves and safety goggles should also be worn. Chloroform is a carcinogen and may damage the liver and kidneys.

Nitric acid is volatile and should be used in a hood. Concentrated acids should be handled with great care; gloves and a face protector should be worn.

Release of Plasmid for Transformation of E. coli *or Yeast*

1. Grow small cultures (at least 1.4 ml) overnight at 30°C in a medium that maintains selection for the plasmid DNA, such as SC – ura.

2. Fill a 1.5-ml microfuge tube with the culture and collect the cells by a 5-second centrifugation in a microfuge.

3. Decant the supernatant and briefly vortex the tube to resuspend the pellet in the residual liquid.

4. Add 0.2 ml of 2% Triton X-100, 1% SDS, 100 mM NaCl, 10 mM Tris-Cl (pH 8), 1 mM Na_2 EDTA. Add 0.2 ml of phenol:chloroform:isoamyl alcohol (25:24:1). Add 0.3 g of acid-washed glass beads.

5. Vortex for 2 minutes.

6. Centrifuge for 5 minutes in a microfuge.

7. Transform 0.2 ml of competent *E. coli* cells with 1–5 µl of the aqueous layer. Transform yeast with 15 µl of the aqueous phase.

Isolation of Genomic DNA for Southern Blot Analysis

1. Grow 10-ml yeast cultures to saturation in YPD at 30°C.

2. Collect the cells by centrifugation for 2 minutes in a clinical centrifuge. Remove the supernatant and resuspend the cells in 0.5 ml of distilled H_2O. Transfer the cells to a 1.5-ml microfuge tube and collect them by centrifugation for 5 seconds in a microfuge.

3. Follow step 3 above.

4. Follow step 4 above.

5. Vortex for 3–4 minutes. Add 0.2 ml of TE (pH 8).

6. Centrifuge for 5 minutes in a microfuge. Transfer the aqueous layer to a fresh tube. Add 1 ml of 100% ethanol. Mix by inversion.

7. Centrifuge for 2 minutes in a microfuge. Discard the supernatant. Resuspend the pellet in 0.4 ml of TE plus 3 µl of a 10 mg/ml solution of RNase A. Incubate for 5 minutes at 37°C. Add 10 µl of 4 M ammonium acetate plus 1 ml of 100% ethanol. Mix by inversion.

8. Centrifuge for 2 minutes in a microfuge. Discard the supernatant. Air-dry the pellet and resuspend in 50 µl of TE. Use 10 µl for each sample to be analyzed by Southern blotting. This is ~2–4 µg of DNA.

MATERIALS AND SOLUTIONS

Medium that maintains selection for the plasmid DNA, such as SC – ura

2% Triton X-100, 1% SDS, 100 mM NaCl, 10 mM Tris-Cl (pH 8), 1 mM Na_2 EDTA (0.4 ml)

Phenol:chloroform:isoamyl alcohol (25:24:1) (0.4 ml)

Glass beads

0.45–0.5-mm beads are available from a variety of suppliers (e.g., Sigma, American QUALEX, Midwest Scientific, Stratagene). Beads can be cleaned by soaking in nitric acid and washing in copious amounts of distilled H_2O. Beads should be dried before use.

YPD

Sterile distilled H_2O

TE (pH 8)

10 mM Tris-Cl (pH 8)

1 mM Na_2 EDTA

100% Ethanol

RNase A stock solution (3 µl)

Dissolve at a concentration of 10 mg/ml in 50 mM potassium acetate (pH 5.5). Boil for 10 minutes. Store frozen at –20°C.

4 M Ammonium acetate (10 µl)

REFERENCES

Hoffman, C.S. and F. Winston. 1987. A ten-minute DNA preparation from yeast efficiently releases autonomous plasmids for transformation of *Escherichia coli*. *Gene* **57:** 267–272.

D. YEAST GENOMIC DNA: GLASS BEAD PREPARATION

1. Grow cells overnight in 5 ml of media (rich or selective) at 30°C in a roller drum.

2. Transfer cultures to 13 x 100-mm glass tubes, and spin down cells in a table-top centrifuge at 1500 rpm for 5 minutes.

3. Wash cells with 3 ml of sterile H_2O, and spin as above.

4. Resuspend in 500 µl of Lysis buffer.

5. Add clean glass beads (about two-thirds of a 1.5-ml Eppendorf tube) and 25 µl of 5 M NaCl.

6. Vortex on highest setting for 1 minute.

7. Spin at 2 krpm for 2 minutes.

8. Transfer the liquid with a P-1000 to a 1.5-ml Eppendorf tube.

9. Add 500 µl of phenol, vortex, and spin for 1 minute. Extract aqueous layer (top) with a P-1000 and transfer to a clean tube. Add 500 µl of SEVAG (24:1 chloroform:isoamyl alcohol), vortex, spin, and extract as above.

10. Add 1 ml of cold 95% ethanol and precipitate for 1 hour at –20°C.

11. Pellet the DNA by spinning for 5 minutes at full speed, pour off the supernatant, and wash with 70% ethanol. Resuspend in 250 µl of TE.

12. Add 25 µl of EDTA-Sark and 5 µl of proteinase K (10 mg/ml). Incubate for 30 minutes at 37°C.

13. Add 250 µl of 5 M NH_4Ac, and repeat steps 9 and 10.

14. Pellet the DNA, wash with 70% ethanol, and resuspend in 100 µl of TE (use ~10 µl/digest).

MATERIALS AND SOLUTIONS

YPD

13 x 100-mm glass tubes

Sterile H_2O

Lysis buffer

0.1 M Tris-Cl (pH 8.0)

50 mM EDTA

1% SDS

5 M NaCl

Glass beads (0.5 mm in diameter; BioSpec Products 11079-105)

SEVAG (24:1 chloroform:isoamyl alcohol)

95% Ethanol

70% Ethanol

EDTA-Sark

0.4 M EDTA (pH 8.0)

2% *N*-lauroylsarcosine (Sarkosyl)

Proteinase K (10 mg/ml)

5 M $NH_4(C_2H_3O_2)$

TE

10 mM Tris-Cl

1 mM EDTA (pH 8.0)

Phenol, equilibrated with H_2O (see Safety Notes in part C)

TECHNIQUES & PROTOCOLS #4

Yeast Protein Extracts

This procedure is useful for making yeast protein extracts for SDS gel electrophoresis and western blotting.

1. The extract is made from 2 OD_{600} units of cells. An easy way to prepare a culture is to inoculate 5 ml of YPD with a very small mass of cells from a plate. This should give an exponential culture (OD_{600} = 0.5–2.0) after overnight growth.

2. Transfer 2 OD_{600} units of cell culture to a 13-mm tube containing 2 ml of 50 mM Tris (pH 7.5), 10 mM NaN_3 on ice. Pellet cells in a swinging-bucket clinical centrifuge.

3. Aspirate supernatant with a 25-gauge needle and suspend cells in 30 µl of ESB.

4. Quickly transfer to a microcentrifuge tube and heat to 100°C for 3 minutes. This heating step should be done rapidly to inactivate proteases. The samples can be stored at –20°C after this.

5. Add about 0.1 g of 0.2-mm glass beads, or pour beads until they reach the top of the liquid. Agitate by vigorous vortex mixing for 2 minutes.

6. Add 70 µl of ESB, vortex briefly, and heat to 100°C for 1 minute.

7. Load 5–20 µl of the extract on an SDS gel.

MATERIALS AND SOLUTIONS

Appropriate growth medium
50 mM Tris (pH 7.5), 10 mM NaN_3
ESB (can be stored at 4°C for months)
  2% SDS
  80 mM Tris (pH 6.8)
  10% Glycerol

1.5 % DTT

0.1 mg/ml Bromophenol blue

Tip:

Quick boiling of the sample is usually sufficient to stabilize proteins against proteolysis. If proteolysis is suspected, PMSF can be added to ESB to reduce proteolysis. In cases of extremely labile proteins, all of the following inhibitors have been found to be beneficial:

Inhibitor	Final concentration	Stock
PMSF	1 mM	100x in isopropanol
pepstatin A	0.7 μg/ml	2000x in methanol
leupeptin	0.5 μg/ml	1000x in H_2O
E64	10 μg/ml	1000x in 50% ethanol
aprotinin	2 μg/ml	5000x in H_2O
α_2 macroglobulin	0.5 U/ml	added directly

For very hydrophobic proteins such as multi-spanning integral membrane proteins, boiling of the sample will cause aggregation. For these proteins, use a protease inhibitor cocktail in ESB and warm the sample to 37°C instead of boiling.

Techniques & Protocols #5

Yeast RNA Isolation

This procedure is designed to yield more than 10 mg of total nucleic acid and 300–400 μg of poly(A)-selected mRNA from each 500-ml culture. The volumes indicated can be applied directly to cultures of 200–500 ml, but should be adjusted for cultures of significantly smaller or larger volumes.

It is convenient to inoculate 200 ml of YPD (or SC) with an appropriate volume of a fresh overnight culture to yield a concentration of 2×10^7–4×10^7 cells/ml the following morning. The cell density should be checked using a hemocytometer (see Appendix G) or Klett (50–100 Klett units for most strains).

Safety Notes

Nitric acid is volatile and should be used in a hood. Concentrated acids should be handled with great care; gloves and a face protector should be worn.

Phenol is highly corrosive and can cause severe burns. Gloves, protective clothing, and safety glasses should be worn when handling it. All manipulations should be carried out in a chemical hood. Any areas of skin that come in contact with phenol should be rinsed with a large volume of water or polyethylene glycol 400 and washed with soap and water; ethanol should not be used.

Diethyl pyrocarbonate is toxic and volatile. Work with open tubes in a hood and wear gloves.

Chloroform is irritating to the skin, eyes, mucus membranes, and upper respiratory tract. It should only be used in a chemical hood. Gloves and safety goggles should also be worn. Chloroform is a carcinogen and may damage the liver and kidneys.

Isolation of Total Yeast RNA

1. Add 50 μg of cycloheximide for each ml of culture. Shake for 15 minutes at 30°C. This step is optional but is supposed to protect mRNA by freezing it in polysomes.

2. To quickly cool the cells for harvesting, fill large centrifuge bottles (Sorvall GS3 or equivalent) halfway with crushed ice, pour the culture over the ice, and shake. Centrifuge at 5000 rpm for 5 minutes at 4°C.

3. Add 11 g of acid-washed glass beads to a 30-ml centrifuge tube. Corex glass tubes work best, but plastic tubes can be used.

4. Add 3 ml of phenol equilibrated with LETS buffer to the glass beads.

5. Resuspend the cell pellet in 2.5 ml of ice-cold LETS buffer and add it to the glass beads/phenol. The meniscus should be just above the surface of the glass beads for the best cell breakage.

6. Vortex at top speed, alternating 30 seconds of vortexing with 30 seconds on ice, for a total of 3 minutes of vortexing. Cell breakage can be checked with a phase-contrast microscope. Broken cells appear as nonrefractile ''ghosts.''

7. When at least 90% of the cells are broken, add 5 ml of ice-cold LETS buffer and vortex briefly. Centrifuge for 5 minutes at 8000 rpm in a Sorvall SS-34 rotor to break the phases. After centrifugation, the phenol layer plus interface should be just below the surface of the beads, allowing easy retrieval of the aqueous phase without disturbing the interface.

8. Transfer the aqueous phase to a clean tube and extract twice with 5 ml of phenol:chloroform:isoamyl alcohol (25:24:1). Avoid transferring the interface. Extract once with chloroform (optional).

9. Add 1/10 volume of 5 M LiCl and precipitate for 3 hours at –20°C. The RNA precipitate may be conveniently stored at –20°C at this point.

Isolation of Poly(A)$^+$ RNA

1. Centrifuge at 10,000 rpm for 10 minutes in a Sorvall SS-34 rotor to pellet the stored RNA. Wash with 80% ethanol. Dry under a vacuum and dissolve in 2.5 ml of distilled H_2O. When dissolved, add 2.5 ml of 2x loading buffer and add SDS to a final concentration of 0.3%.

2. Heat for 5 minutes at 65°C.

3. Load an oligo(dT) column and wash three times with 1x loading buffer.

4. Elute the poly(A)+ RNA with 10 mM HEPES (pH 7); add RNase inhibitor if desired.

5. Determine the RNA concentration by measuring the OD_{260} of a 1:10 dilution of the poly(A)+ RNA in distilled H_2O. Compare this with a 1:10 dilution of 10 mM HEPES (pH 7) in H_2O (i.e., compare with 1 mM HEPES).

6. Precipitate the RNA by adding 1/10 volume of 3 M sodium acetate and 2.5 volumes of 100% ethanol. Mix by inversion and incubate for 30 minutes at –70°C. Centrifuge for 10 minutes in a microfuge. Remove the supernatant and dissolve the RNA pellet in distilled H_2O to make a final concentration of 1 mg/ml. Either the dissolved RNA or the ethanol precipitate can be stored indefinitely at –70°C.

MATERIALS

Cycloheximide (50 µg/ml of culture) (optional)
Large centrifuge bottles (Sorvall GS3 or equivalent)
Glass beads

0.45–0.5-mm beads are available from a variety of suppliers (e.g., Sigma, American QUALEX, Midwest Scientific, Stratagene). Beads can be cleaned by soaking in nitric acid and washing in copious amounts of distilled H_2O. Beads should be dried before use.

Centrifuge tubes (Corex glass or plastic; 30-ml)
Phenol equilibrated with LETS buffer (3 ml)
LETS buffer

- 0.1 M Lithium chloride (LiCl)
- 0.01 M Na_2 EDTA
- 0.01 M Tris-Cl (pH 7.4)
- 0.2% SDS
- 0.1% Diethyl pyrocarbonate (optional)

Phenol:chloroform:isoamyl alcohol (25:24:1)
Chloroform (optional)

5 M LiCl

80% Ethanol

Sterile distilled H_2O

1x Loading buffer

- 0.5 M NaCl
- 0.01 M HEPES (pH 7)

SDS

Oligo(dT) column

10 mM HEPES (pH 7)

RNase inhibitor (optional)

3 M Sodium acetate

100% Ethanol

TECHNIQUES & PROTOCOLS #6

Hydroxylamine Mutagenesis of Plasmid DNA

Safety Notes

Hydroxylamine is a mutagen and should be handled carefully. Gloves should be worn when handling it.

Solid NaOH is caustic and should be handled with great care; gloves and face protector should be worn.

1. Prepare the hydroxylamine solution just before use and store on ice until needed.

2. Add 10 μg of CsCl-purified plasmid DNA to 500 μl of hydroxylamine solution in a microfuge tube.

3. Incubate for 20 hours at 37°C.

4. Stop the reaction by adding 10 μl of 5 M NaCl, 50 μl of 1 mg/ml BSA, and 1 ml of 100% ethanol; precipitate the DNA for 10 minutes at –70°C.

5. Centrifuge the precipitated DNA in a microfuge for 10 minutes. Carefully remove all of the supernatant.

6. Resuspend the DNA in 100 μl of TE (pH 8). Add 10 μl of 3 M sodium acetate and 250 μl of 100% ethanol; precipitate the DNA for 10 minutes at –70°C and centrifuge as in step 5.

7. Allow the pellet to air-dry and then resuspend it in 100 μl of TE (pH 8).

8. The DNA can be used directly for transformation of either *E. coli* or yeast. The transformation frequency in yeast with the mutagenized DNA is reduced only approximately threefold relative to unmutagenized DNA. The formation of Ura^- plasmids in *E. coli* can be monitored. The *E. coli* strain can be ung^+, but the transformation frequency will be reduced 10- to 100-fold. When the *URA3* gene

makes up approximately 10% of the plasmid DNA and an *E. coli ung*$^+$ *pyrF*$^-$ strain (DB6507) is being transformed, approximately 4% Ura$^-$ colonies can be expected. Transforming yeast with the same stock of mutagenized DNA gives approximately 1% loss-of-function mutations (''knockouts'') in my favorite gene. Approximately 10% of the mutants are temperature-sensitive.

MATERIALS AND SOLUTIONS

Hydroxylamine solution

0.35 g Hydroxylamine HCl

0.09 g NaOH

5 ml Distilled H_2O (ice-cold)

Dissolve the solids in the H_2O. The pH should be ~7. Prepare just before use and store on ice until needed.

CsCl-purified plasmid DNA (10 μg)

5 M NaCl

1 mg/ml Bovine serum albumin (BSA)

100% Ethanol

TE (pH 8)

10 mM Tris-Cl (pH 8)

1 mM Na_2 EDTA

3 M Sodium acetate

TECHNIQUES & PROTOCOLS #7

Assay of β-Galactosidase in Yeast

There are two basic methods for the in vitro assay of β-galactosidase from yeast. They differ mainly in the method of preparing the material for assay. In the first method (Rose and Botstein 1983), a crude extract is prepared, and the activity is normalized to the amount of protein assayed. In the second method (Guarente 1983), the cells are permeabilized to allow the substrate to enter the cells, and the activity is normalized to the number of cells assayed. The former method is preferable when comparing cells that are grown under very different conditions or that have different genetic backgrounds. The latter method is adapted from the assay for *E. coli* and is particularly suited for changing levels of activity within a single strain.

Method I: Assay of Crude Extracts

Safety Notes

Nitric acid is volatile and should be used in a hood. Concentrated acids should be handled with great care; gloves and a face protector should be worn.

Phenylmethylsulfonyl fluoride (PMSF) is extremely destructive to the mucous membranes of the respiratory tract, the eyes, and the skin. It may be fatal if inhaled, swallowed, or absorbed through the skin. In case of contact, immediately flush eyes or skin with copious amounts of water and discard contaminated clothing.

Chloroform is irritating to the skin, eyes, mucus membranes, and upper respiratory tract. It should only be used in a chemical hood. Gloves and safety goggles should also be worn. Chloroform is a carcinogen and may damage the liver and kidneys.

1. Grow a 5-ml culture of cells to a concentration of 1×10^7–2×10^7 cells/ml in an appropriate liquid medium at an appropriate temperature (usually 30°C). If the hybrid gene is expressed from an autonomous plasmid, use an appropriate medium to select for the presence of the plasmid.

2. Chill the cells on ice and harvest by centrifugation (2000 rpm for 5 minutes in a clinical centrifuge is adequate).

Keep the cells on ice from this point on.

3. Discard the supernatant. Resuspend the cells in 250 μl of breaking buffer. The cells can now be frozen at –20°C and assayed at a later date.

All of the following steps can be performed in a 1.5-ml microfuge tube.

4. If the cells were frozen, thaw them on ice. Add glass beads until the beads reach a level just below the meniscus of the liquid. Add 12.5 μl of PMSF stock solution.

5. Vortex six times at top speed in 15-second bursts. Chill on ice between bursts.

6. Add 250 μl of breaking buffer and mix well. Withdraw the liquid extract after plunging the tip of a 1000-μl pipettor to the bottom of the tube.

7. Clarify the extract by centrifuging for 15 minutes in a microfuge. If the activity is in the particulate fraction, the unclarified supernatant can be used and the assay mixture may be clarified later in step 8.

8. To perform the assay:

   a. Add 10–100 μl of extract to 0.9 ml of Z buffer. Adjust the volume to 1 ml with breaking buffer.

   b. Incubate the mixture in a water bath at 28°C for 5 minutes.

   c. Initiate the reaction by adding 0.2 ml of ONPG stock solution. Note precisely the time that the addition is made. Incubate at 28°C until the mixture has acquired a pale yellow color.

   d. Terminate the reaction by adding 0.5 ml of Na_2CO_3 stock solution. Note precisely the time that the reaction is terminated. Measure the optical density at 420 nm.

9. Measure the protein concentration in the extract using the dye-binding assay of Bradford (1976):

a. Dilute the Bradford reagent fivefold in distilled H_2O. Filter the diluted reagent through Whatman 540 paper (or equivalent).

b. Add 10–20 µl of the extract to 1 ml of the diluted reagent and mix. Measure the blue color formed at 595 nm. Use disposable plastic cuvettes to prevent the formation of a blue film.

c. Prepare a standard curve using several dilutions (0.1–1 mg/ml) of BSA dissolved in breaking buffer.

Typical extracts prepared in this fashion contain 0.5–1 mg/ml of protein.

10. Express the specific activity of the extract according to the following formula:

$$\frac{OD_{420} \times 1.7}{0.0045 \times \text{protein} \times \text{extract volume} \times \text{time}}$$

OD_{420} is the optical density of the product, *o*-nitrophenol, at 420 nm. The factor 1.7 corrects for the reaction volume. The factor 0.0045 is the optical density of a 1 nmole/ml solution of *o*-nitrophenol. Protein concentration is expressed as mg/ml. Extract volume is the volume assayed in ml. Time is in minutes. Specific activity is expressed as nmoles/minute/mg protein.

Method II: Permeabilized Cell Assay

1. Grow the cells as above. Measure the OD_{600} of the culture and harvest 1×10^6–1×10^7 cells by centrifugation as above.

2. Discard the supernatant. Resuspend the cells in 1 ml of Z buffer.

3. Add 3 drops of chloroform and 2 drops of 0.1% SDS. Vortex at top speed for 10 seconds.

4. Preincubate the samples for 5 minutes at 28°C. Start the reaction by adding 0.2 ml of ONPG as above.

5. Stop the reaction by adding 0.5 ml of Na_2CO_3 stock solution when the sample in the tube has developed a pale yellow color. Note the amount of time elapsed during the assay. Remove the cell debris by centrifuging for 10 minutes in a microfuge and then discarding the pellet.

6. Measure the OD_{420} of the reactions.

7. Express the activity as β-galactosidase units:

$$\frac{OD_{420}}{OD_{600} \text{ of assayed culture} \times \text{volume assayed} \times \text{time}}$$

   OD_{420} is the optical density of the product, *o*-nitrophenol. OD_{600} is the optical density of the culture at the time of assay. Volume is the amount of the culture used in the assay in ml. Time is in minutes.

MATERIALS AND SOLUTIONS

Appropriate liquid medium

Breaking buffer

  100 mM Tris-Cl (pH 8)

  1 mM Dithiothreitol

  20% Glycerol

Glass beads

  0.45–0.5-mm beads are available from a variety of suppliers (e.g., Sigma, American QUALEX, Midwest Scientific, Stratagene). Beads can be cleaned by soaking them in nitric acid and then washing them in copious amounts of distilled H_2O. Beads should be dried before use.

PMSF (Sigma P 7626) stock solution

  40 mM in 100% isopropanol. Store at –20°C.

Z buffer (Miller 1972)

  16.1 g $Na_2HPO_4 \cdot 7H_2O$

  5.5 g $NaH_2PO_4 \cdot H_2O$

  0.75 g KCl

0.246 g $MgSO_4 \cdot 7H_2O$

2.7 ml β-Mercaptoethanol

Distilled H_2O to make a final volume of 1 liter

Adjust the pH to 7. Store at 4°C.

ONPG (*o*-nitrophenyl-β-D-galactoside) stock solution

4 mg/ml in Z buffer. Store at –20°C.

Na_2CO_3 stock solution

1 M in distilled H_2O

Bradford reagent (Bio-Rad)

Distilled H_2O

Whatman 540 paper or equivalent

Disposable plastic cuvettes

0.1–1 mg/ml Bovine serum albumin (BSA) in breaking buffer

Chloroform

0.1% SDS

REFERENCES

Bradford, M.M. 1976. A dye binding assay for protein. *Anal. Biochem.* **72:** 248–254.

Guarente, L. 1983. Yeast promoters and *lacZ* fusions designed to study expression of cloned genes in yeast. *Methods Enzymol.* **101:** 181–191.

Miller, J.H. 1972. *Experiments in molecular genetics.* Cold Spring Harbor Laboratory, Cold Spring Harbor, New York.

Rose, M. and D. Botstein. 1983. Construction and use of gene fusions *lacZ* (β-galactosidase) which are expressed in yeast. *Methods Enzymol.* **101:** 167–180.

Techniques & Protocols #8

Plate Assay for Carboxypeptidase Y

(Adapted from Jones [1991] *Methods in Enzymology* **194:** 428–453 [pp. 436–439].)

1. Grow colonies or patches on YPD plates (3 days for colonies and 1 day for patches is usually sufficient).

2. Carefully pour overlay solution over the surface of the plate to completely cover the cells.

3. After the agar hardens in 5–10 minutes, carefully flood the surface with *fresh* Fast Garnet GBC solution.

4. Allow color to develop for 5 minutes. Wild-type strains will turn red, whereas carboxypeptidase Y-negative strains will appear yellow or pink.

5. Decant Fast Garnet solution to best observe developed color.

MATERIALS AND SOLUTIONS

Overlay solution for one plate

In glass or polypropylene tube, mix 2.5 ml of 1 mg/ml *N*-acetyl-DL-phenylalanine β-naphthyl ester in dimethylformamide with 4 ml of 0.6% molten agar. Hold at 50°C.

Fast Garnet GBC solution (use 5 ml/plate)

5 mg/ml Fast Garnet GBC (sulfate salt; Sigma F 8761) in 0.1 M Tris HCl (pH 7.4)

Dimethylformamide permeabilizes the cells. Carboxypeptidase Y in the cells cleaves the ester linkage in *N*-acetyl-DL-phenylalanine β-naphthyl ester. Free β-naphthol then reacts with the diazonium salt Fast Garnet GBC to produce an insoluble red dye.

REFERENCE

Jones, E.W. 1991. Tackling the protease problem in *Saccharomyces cerevisiae*. *Methods Enzymol.* **194:** 428–453.

TECHNIQUES & PROTOCOLS #9

Random Spore Analysis

I. Sporulation

1. Patch out a single colony of the diploid to be sporulated onto YPD. It is best to spread the cells as thinly as possible. Incubate for 12–16 hours at 30°C. Growth for more than 16 hours will dramatically decrease the efficiency of sporulation.

2. In the morning, with a sterile dowel transfer a matchhead quantity of cells to a test tube containing 2.5 ml of sporulation medium and place on a rotor at 25°C.

3. Monitor the extent of sporulation by light microscopy. It can take from 2–10 days for more than 5% of cells to sporulate.

II. Random Spores

4. Transfer 1 ml of sporulated culture to a 15-ml conical polystyrene tube and collect the cells by centrifugation for 5 minutes in the clinical centrifuge.

5. Remove the supernatant completely and resuspend cells in 0.2 ml of sterile H_2O and 5 μl of β-glucuronidase (~500 units).

6. Incubate on a rotor for 1 hour at 30°C.

7. Add 0.1 ml (~0.15 g) sterile 0.5-mm glass beads. Incubate on a rotor for 1 hour at 30°C.

8. Add 1 ml of sterile H_2O.

9. Vortex 1–2 minutes and check by light microscopy for complete disruption of asci.

10. Add 4 ml of sterile H_2O.

11. Make dilutions of 10^{-1} to 10^{-3} in sterile H_2O. Plate 200 µl onto SC-arg with 60 µg/ml canavanine.

MATERIALS AND SOLUTIONS

Sporulation medium

1% KOAc

0.025% Glucose

β-Glucuronidase (Sigma G 7770)

Sterile 0.5-mm glass beads

SC -arg with 60 µg/ml canavanine plates

Techniques & Protocols #10

Yeast Vital Stains

Nuclear and Mitochondrial DNA

Safety Note

DAPI is a possible carcinogen. It may be harmful if inhaled, swallowed, or absorbed through the skin. It may also cause irritation. Wear gloves, face mask, and safety glasses and do not breathe the dust.

1. Pellet ~10^7 cells in a microfuge tube (5 second pulse) and resuspend in 70% ethanol.

2. Fix for 5 minutes or more and wash twice with H_2O.

3. Suspend cells in a small volume of 50 ng/ml DAPI (4′6,-diamidino-2-phenylindole; Sigma D 9542 or Accurate Chemical and Scientific Corp.) in mounting medium. A stock of 1 mg/ml in H_2O can be stored at –20°C.

4. Observe with UV filter set.

Cells fixed in formaldehyde can also be stained with DAPI in mounting medium. DAPI can also be used as a vital stain of cells in growth medium at a concentration of ~1 μg/ml but the background staining of the cell bodies is higher than with fixed cells.

Mitochondria

1. To ~10^7 cells in growth medium, add 100 ng/ml $DiOC_6$ (3,3′-dihexyloxacarbocyanine iodide; Sigma D 3652 or Molecular Probes) (dilute $1/10^4$ from 1 mg/ml stock in ethanol which is stable for months in the dark at –20°C).

2. Incubate for 5 minutes or more and observe with fluorescein filter set.

The concentration of $DiOC_6$ may need to be optimized—at ~1 µg/ml all membranes appear to be stained.

Vacuole

1. Pellet ~10^7 cells in a microfuge tube (5-second pulse) and resuspend in YPD with 50 mM sodium citrate (pH 4.0).

2. Add CDCFDA (carboxy-2′,7′-dichloro-fluorescein diacetate; Molecular Probes) to 10 µM (1/1000 dilution of a 10 mM stock in dimethylformamide which is stable for months at –20°C).

3. Incubate for 10 minutes or more and observe with fluorescein filter set.

Bud Scars and Chitin

1. To ~10^7 cells in growth medium, add 100 µg/ml Calcofluor (Fluorescent brightener 28; Sigma F 3543) (dilute 1/10 from 1 mg/ml stock in H_2O which is stable for weeks in the dark at –20°).

2. Incubate 5 minutes or more, wash twice with H_2O, and observe with UV filter set.

TECHNIQUES & PROTOCOLS #11

Yeast Immunofluorescence

Preparation of Cells

Safety Note

Formaldehyde is toxic and is a carcinogen. It is readily absorbed through the skin and is irritating to the eyes, skin, mucus membranes, and upper respiratory tract. Wear gloves and safety glasses and always work in a chemical hood.

1. Grow a 5-ml culture to early exponential phase (10^6–10^7 cells/ml).

2. Fix cells by adding 1/10 volume formaldehyde directly to medium (total formaldehyde concentration 3.7%; standard stock solution is 37%).

3. Incubate cells in formaldehyde for at least 1 hour.

4. Pellet cells and wash once with 0.1 M potassium phosphate (pH 7.5).

5. Resuspend cells in 1 ml of 50 mg/ml Zymolyase 100T or ~50 units/ml lyticase in 0.1 M potassium phosphate (pH 7.5) with 2 ml/ml b-mercaptoethanol.

6. Incubate for 30 minutes at 30°C. Check efficiency of spheroplasting by phase contrast microscopy. The cells should be a dark, translucent gray. Bright (refractile) cells are insufficiently digested. Ghosts (pale gray with little if any internal structure) have been overdigested.

7. Pellet and resuspend in 1 ml of PBS.

Staining

Safety Note

DAPI is a possible carcinogen. It may be harmful if it is inhaled, swallowed, or absorbed through the skin. It may also cause irritation. Wear gloves, face mask, and safety glasses, and do not breathe in the dust.

1. Prepare Teflon-masked slides (10-well multiwell slides) by putting 10 ml of 1 mg/ml polylysine (size 400,000) onto each well. Wash slide with distilled H_2O and let dry.

2. Place 10 ml of fixed cells onto each well. After a few minutes, aspirate and wash three times with PBS. Check the slide in the microscope to ensure the cells are at a suitable density and not clumped.

3. Optional step (not necessary for the antibodies used in this course, but recommended):

   Dunk the slide in cold methanol (-20°C) for 6 minutes and then into cold acetone (-20°C) for 30 seconds. This treatment results in flatter cells, which can aid visualization of the cytoskeleton. For some antibody-antigen combinations, this step might be necessary for reactivity. Rehydrate wells with PBS.

4. Place 15 ml of blocking buffer consisting of PBS + 3% BSA (bovine serum albumin) on wells. Incubate for 30 minutes in a humid chamber such as a Petri dish with a wet Kimwipe in it. It is important to not let slide wells dry out from this point on. (The BSA reduces nonspecific antibody binding to the slide by blocking protein binding sites).

5. Aspirate off the blocking buffer and wash two times with PBS + 3% BSA.

6. Add 10-15 ml of primary antibody in PBS + 3% BSA to the wells and incubate for 1 hour in a humid chamber. Use affinity-purified antibodies or monoclonals, and if possible, use an isogenic control that lacks the antigen of interest. It is also helpful to do a control without primary antibody.

7. Aspirate off primary antibody and wash three times with PBS + 3% BSA.

8. Repeat steps 6 and 7 with fluorescent secondary antibody (incubate in the dark).

9. Aspirate off secondary antibody and wash three times with PBS + 3% BSA.

10. Add 10-15 ml of 1 μg/ml DAPI to wells, incubate for about 1 minute, and wash three times with PBS.

11. Aspirate the PBS, and add a small drop of mounting medium to each well. Put on the cover slip, trying to avoid trapping air bubbles in the wells. Remove excess mounting solution with a Kimwipe, taking care not to move the coverslip. Seal with clear nail polish and store at $-20^{o}C$.

MATERIALS AND SOLUTIONS

Zymolyase® 100T (120493-1, Seikagaku America Inc.)

Mounting medium

Dissolve 50 mg *p*-phenylenediamine in 5 ml of PBS (see below) and adjust to pH 9. Add 45 ml of glycerol and stir until homogeneous. Store in aliquots at $-70^{o}C$ in the dark. (Modified from Johnson and Nogueira Araujo 1981.)

If DNA-staining is also required, 4'6,-diamidino-2-phenylindole (DAPI; Sigma D 9542 or Accurate Chemical and Scientific Corp.) is included in the mounting medium (Williamson and Fennell 1975). To 50 ml of mounting medium, add 2.5 ml of fresh DAPI solution (1 mg DAPI/ml H_20).

PBS

A 20x stock contains per liter:

160 g NaCl

4 g KCl

22.8 g Na_2HPO_4

4 g KH_2PO_4

Adjust to pH 7.3 with 10 N NaOH.

REFERENCES

Johnson, G.D. and G.M. Nogueira Araujo. 1981. A simple method of reducing the fading of immunofluorescence during microscopy. *J. Immunol. Methods* **43:** 349–350.

Williamson, D.H. and D.J. Fennell. 1975. The use of fluorescent DNA-binding agent for detecting and separating yeast mitochondrial DNA. *Methods Cell Biol.* **12:** 335–351.

Techniques & Protocols #12

Actin Staining in Fixed Cells

Phalloidin binds specifically to F-actin and fluorescent tagged phalloidin stains the actin skeleton in cells in a manner that is very close to the staining pattern seen using anti-actin antibody.

Safety Note

Formaldehyde is toxic and is a carcinogen. It is readily absorbed through the skin and is irritating to the eyes, skin, mucus membranes, and upper respiratory tract. Wear gloves and safety glasses and always work in a chemical hood.

1. Grow cells to exponential phase (~10^7 cells/ml).

2. Fix cells in a microfuge tube by adding 0.1 ml formaldehyde directly to 1 ml of culture medium (formaldehyde concentration 3.7%; standard stock solution is 37%).

3. Incubate cells in formaldehyde for 30 minutes or more.

4. Wash two times with PBS; pelleting with 5-second pulses in microfuge.

5. Suspend cells in 50 µl of PBS and add 5 U of rhodamine or fluorescein-conjugated phalloidin (normally this would be 25 µl of a 200 U/ml stock in methanol which is stable for months at –20°C).

6. Wash three times with PBS and resuspend in a small volume of mounting medium.

7. Observe with either rhodamine or fluorescein filter sets.

TECHNIQUES & PROTOCOLS #13

PCR Protocol for PCR-mediated Gene Disruption

(Adapted from Brachmann et al. [1997] *Yeast*. [in press])

1. Reaction mix:

   5 µl of 10X *Taq* Buffer
   5 µl of 25 mM $MgCl_2$
   2 µl of 10 mM dNTP's
   10–100 ng of template DNA
   25 pmols of each primer
   0.5 µl of *Taq* polymerase (2 U)

   ____

   ==> H_2O to 50 µl total volume

2. PCR cycle profile:

   94°C 5 minutes

   ____

   94°C 1 minute
   55°C 1 minute
   72°C 2 minutes
   ==> 10 cycles

   ____

   94°C 1 minute
   65°C 1 minute
   72°C 2 minutes
   ==> 20 cycles

   ____

   72°C 10 minutes

Note: Typically, the entire reaction can be used for a transformation, although purifying the amplified DNA product increases transformation efficiency dramatically. We have success using Wizard™PCR Preps DNA Purification System (Promega Corp. A 7170).

MATERIALS AND SOLUTIONS

10X *Taq* Buffer

- 0.5 M KCl
- 100 mM Tris-Cl (pH 8.5)
- 0.1% Triton X-100

25 mM $MgCl_2$

10 mM dNTPs

pRS40X template DNA (mini-prep DNA works well)

Taq polymerase

Two gene-specific DNA primers

One oligonucleotide should consist of 40 nucleotides of gene-specific sequence for one end of the targeted region at the 5' end followed by:

5'-CTGTGCGGTATTTCACACCG-3' (left primer),

and another 40 nucleotides homologous to the other side of the targeted region at the 5' end followed by:

5' AGATTGTACTGAGAGTGCAC-3' (right primer).

The primers are then used to amplify any auxotrophic marker from a pRS40X or pRS30X integrating plasmid (Sikorski and Hieter 1989; Brachmann et al. 1997).

REFERENCES

Brachmann, C.B., A. Davies, G.J. Cost, E. Caputo, J. Li, P. Hieter, and J.D. Boeke. 1997. Designer deletion strains derived from *Saccharomyces cerevisiae* S288C: A useful set of strains and plasmids for PCR-mediated gene disruption and other applications. *Yeast* (in press).

Sikorski, R.S. and P. Hieter. 1989. A system of shuttle vectors and yeast host strains designed for efficient manipulation of DNA in *Saccharomyces cerevisiae*. *Genetics* **122:** 19-27.

Techniques & Protocols #14

Yeast Colony PCR Protocol

1. Combine reaction mix on ice:

    2 µl of 10x Colony PCR Buffer
    1.2 µl of 25 mM $MgCl_2$
    0.4 µl of 10 mM dNTPs
    10 pmols of each primer
    0.2 µl *Taq* polymerase (5 U)

    ==> H_2O to 20 µl

2. Using a pipette tip, add a very small amount of cells (~0.25 µl) into PCR reaction mix (20 µl reaction).

3. PCR cycle profile:

    94° 4 minutes

    94°C 1 minute
    55°C 1 minute
    72°C 2 minutes
    ==> 35 cycles

    72°C 10 minutes

4. Load entire sample on agarose gel.

MATERIALS AND SOLUTIONS

10x Colony PCR Buffer

0.125 M Tris-HCl (pH 8.5)
0.56 M KCl

25 mM $MgCl_2$

10 mM dNTPs

Taq polymerase

Two gene-specific DNA primers

Each oligonucleotide should be 25 nucleotides long and specific for either side of the region of interest to be amplified.

Note: The elongation times (at 72°C) work well for amplification of loci ≤1.5 kbp in size. These conditions may need to be modified for amplification of longer regions.

Appendix A

Media

Media for petri plates are prepared in 2-liter flasks, with each flask containing 1 liter of medium, which is sufficient for 30–40 plates. Unless otherwise stated, all components are autoclaved together for 15 minutes at 250°F (121°C) and 15 lb/sq. in. of pressure. Longer autoclaving of minimal media leads to hydrolysis of the agar, caramelized glucose, and mushy plates. If larger volumes are to be prepared, autoclave the salts, glucose, and agar separately for longer periods of time. The plates should be allowed to dry at room temperature for 2–3 days after pouring. The plates can be stored in sealed plastic bags for over 3 months. The agar is omitted for liquid media. (For convenience, the final concentration of each component in the medium is listed in parentheses below.)

YPD (YEPD)

YPD is a complex medium for routine growth.

Bacto-yeast extract (1%)	10 g
Bacto-peptone (2%)	20 g
Glucose (2%)	20 g
Bacto-agar (2%)	20 g
Distilled H_2O	1000 ml

YPG (YEPG or YEP-Glycerol)

YPG is a complex medium containing a nonfermentable carbon source (glycerol) that does not support the growth of ρ^- or *pet* mutants.

Bacto-yeast extract (1%)	10 g
Bacto-peptone (2%)	20 g
Glycerol (3% [v/v])	30 ml
Bacto-agar (2%)	20 g
Distilled H_2O	970 ml

YPDG

YPDG is a complex medium that can be used to differentiate between ρ^+ and ρ^- colonies.

Combine:

Bacto-agar (2%)	20 g
Bacto-yeast extract (1%)	10 g
Bacto-peptone (2%)	20 g
Distilled H_2O	900 ml

After autoclaving the above ingredients together, add sterile:

30% Glycerol (3%)	100 ml
20% Glucose (0.1%)	5 ml

YPAD (SLANT MEDIUM)

YPAD is a complex medium used for the preparation of slants. The adenine is added to inhibit the reversion of *ade1* and *ade2* mutants.

Bacto-yeast extract (1%)	10 g
Bacto-peptone (2%)	20 g
Glucose (2%)	20 g
Adenine sulfate (0.004%)	40 mg
Bacto-agar (2%)	20 g
Distilled H_2O	1000 ml

Dissolve the medium in a boiling-water bath. Dispense 1.5-ml portions with an automatic pipettor into 1-dram vials. Screw on the caps loosely and autoclave the vials. After autoclaving, incline the rack so that the agar is just below the neck of the vial. Tighten the caps after 1–2 days.

SYNTHETIC DEXTROSE MINIMAL MEDIUM (SD)

SD is a synthetic minimal medium containing salts, trace elements, vitamins, a nitrogen source (Bacto-yeast nitrogen base without amino acids), and glucose.

Bacto-yeast nitrogen base without amino acids (0.67%)	6.7 g
Glucose (2%)	20 g
Bacto-agar (2%)	20 g
Distilled H_2O	1000 ml

SUPPLEMENTED MINIMAL MEDIUM (SMM)

SMM is SD to which various growth supplements have been added. The specific constituents in SMM are defined in the Materials section of each experiment, where applicable. It is convenient to prepare sterile stock solutions by autoclaving for 15 minutes at 250°F (121°C). These solutions can then be stored for extensive periods. Some should be stored at room temperature in order to prevent precipitation, whereas the other solutions may be refrigerated. Wherever applicable, HCl salts of amino acids are preferred.

The medium should be prepared by adding the appropriate volumes of the stock solutions to the ingredients of SD medium and then adjusting the total volume to 1 liter with distilled H_2O. Threonine and aspartic acid solutions should be added separately to the medium after it is autoclaved.

Alternatively, it is often more convenient to prepare the medium by spreading a small quantity of the supplement(s) on the surface of an SD plate. The solution(s) should then be allowed to dry thoroughly onto the plate before inoculating it with yeast strains.

Given below are the concentrations of the stock solutions, the volume of stock solution necessary for mixing a liter of medium, and the volume of stock solution to spread on SD plates. The final concentration of each constituent in SMM is also given.

Constituent	Stock concentration (g/100 ml)	Volume of stock for 1 liter of medium (ml)	Final concentration in medium (mg/liter)	Volume of stock to spread on plate (ml)
Adenine sulfate	0.2[a]	10	20	0.2
Uracil	0.2[a]	10	20	0.2
L-Tryptophan	1	2	20	0.1
L-Histidine HCl	1	2	20	0.1
L-Arginine HCl	1	2	20	0.1
L-Methionine	1	2	20	0.1
L-Tyrosine	0.2	15	30	0.2
L-Leucine	1	10	100	0.1
L-Isoleucine	1	3	30	0.1
L-Lysine HCl	1	3	30	0.1
L-Phenylalanine	1[a]	5	50	0.1
L-Glutamic acid	1[a]	10	100	0.2
L-Aspartic acid	1[a,b]	10	100	0.2
L-Valine	3	5	150	0.1
L-Threonine	4[a,b]	5	200	0.1
L-Serine	8	5	400	0.1

[a]Store at room temperature.
[b]Add after autoclaving the medium.

SYNTHETIC COMPLETE (SC) AND DROP-OUT MEDIA

In order to test the growth requirements of strains, it is useful to have media in which each of the commonly encountered auxotrophies is supplemented except the one of interest (drop-out media). Dry growth supplements are stored premixed.

SC is a medium in which the drop-out mix contains all possible supplements (i.e., nothing is ''dropped out'').

Bacto-yeast nitrogen base without amino acids (0.67%)	6.7 g
Glucose (2%)	20 g
Bacto-agar (2%)	20 g
Drop-out mix (0.2%)	2 g
Distilled H_2O	1000 ml

Drop-out mix:

Drop-out mix is a combination of the following ingredients minus the appropriate supplement. It should be mixed very thoroughly by turning end-over-end for at least 15 minutes; adding a couple of clean marbles helps.

Adenine	0.5 g	Leucine	10.0 g
Alanine	2.0 g	Lysine	2.0 g
Arginine	2.0 g	Methionine	2.0 g
Asparagine	2.0 g	*para*-Aminobenzoic acid	0.2 g
Aspartic acid	2.0 g	Phenylalanine	2.0 g
Cysteine	2.0 g	Proline	2.0 g
Glutamine	2.0 g	Serine	2.0 g
Glutamic acid	2.0 g	Threonine	2.0 g
Glycine	2.0 g	Tryptophan	2.0 g
Histidine	2.0 g	Tyrosine	2.0 g
Inositol	2.0 g	Uracil	2.0 g
Isoleucine	2.0 g	Valine	2.0 g

HARTWELL'S COMPLETE (HC) MEDIA

HC media, used in Lee Hartwell's lab, is used in the same way as SC media (see above); however, it uses different combinations of supplements for growth. This difference supports better growth of some strains. In addition, it is presented as a series of stock solutions, which gives very reproducible results between batches of media, and with a preference for easily making the six most common drop-out media.
The stock solutions are:

10x HC (6 Dropout amino acid liquid):

Methionine	0.8 g
Tyrosine	2.4 g
Isoleucine	3.2 g
Phenylalanine	2.0 g
Glutamic acid	4.0 g
Threonine	8.0 g
Aspartic acid	4.0 g
Valine	6.0 g
Serine	16.0 g
Arginine	0.8 g

in a final volume of 4 liters; autoclave

10x YNB:

Yeast nitrogen base	58 g
(w/o amino acids and w/o ammonium sulfate)	
Ammonium sulfate	200 g
in a final volume of 4 liters; autoclave	

Amino acid solutions (note that these are NOT 10x solutions; see below)

Sterilize each solution by autoclaving. Keep the Tryptophan in a light-sensitive bottle.

Uracil	1 g/liter
Adenine	1 g/liter
Lysine	10 g/liter
Tryptophan	10 g/liter
Leucine	20 g/liter
Histidine	10 g/liter

Recipe for Hartwell Complete Plates:

20 g of Agar in 619 ml of distilled H_2O in a 2-liter flask. Autoclave 20 minutes. Then add the following:

20% Glucose (sterile)	100 ml
10x YNB solution	100 ml
10x HC dropout 6 amino acid solution	100 ml
Uracil solution	35 ml
Adenine solution	20 ml
Lysine solution	12 ml
Tryptophan solution	8 ml
Leucine solution	4 ml
Histidine solution	2 ml

MAL INDICATOR MEDIUM

MAL indicator medium is a fermentation-indicator medium used to distinguish strains that ferment or do not ferment maltose. Due to the pH change, the maltose-fermenting strains will change the indicator yellow.

Bacto-yeast extract (1%)	10 g
Bacto-peptone (2%)	20 g

Maltose (2%)	20 g
Bromcresol purple solution (0.4% stock solution)	9 ml
Bacto-agar (2%)	20 g
Distilled H_2O	1000 ml

0.4% Bromcresol purple solution:

Bromcresol purple	200 mg
100% Ethanol	50 ml

GAL INDICATOR MEDIUM

GAL indicator medium is used for scoring the ability to ferment galactose.

Bacto-yeast extract (1%)	10 g
Peptone (2%)	20 g
Bacto-agar (2%)	20 g
Bromthymol blue solution (4 mg/ml stock solution)	20 ml
Distilled H_2O	880 ml

After autoclaving, add 100 ml of a filter-sterilized (0.2-μm filter) 20% galactose solution.

Bromthymol blue solution:

Bromthymol blue	400 mg
Distilled H_2O	100 ml

X-GAL INDICATOR PLATES FOR YEAST

5-Bromo-4-chloro-3-indolyl-β-D-galactoside (X-gal) does not work for yeast at the normal acidic pH of SD medium; therefore, a neutral pH medium is used. This is clearly a trade-off as many yeast strains will not grow well at this pH. For a first attempt at assessing β-galactosidase expression, this medium is worth a shot.

For 1 liter of X-gal indicator plates:

Solution I:

Mix:

10x Phosphate-buffer stock solution	100 ml
1000x Mineral stock solution	1 ml
Drop-out mix	2 g

Adjust the volume to 450 ml with distilled H_2O if the medium is to contain glucose or to 400 ml if it is to contain galactose.

Solution II:

Mix in a 2-liter flask:

Bacto-agar	20 g
Distilled H_2O	500 ml

Autoclave the solutions separately. After cooling to below 65°C, add the following to Solution I:

Glucose or other sugar to a final concentration of 2%	
X-gal (20 mg/ml dissolved in dimethylformamide)	2 ml
100x Vitamin stock solution	10 ml
Any other heat-sensitive supplements	

Mix the solutions together and pour ~30 ml/plate.

10x Phosphate-buffer stock solution:

KH_2PO_4 (1 M)	136.1 g
$(NH_4)_2SO_4$ (0.15 M)	19.8 g
KOH (0.75 N)	42.1 g
Distilled H_2O	1000 ml

Adjust the pH to 7 and autoclave. **Safety Note:** Solid KOH is caustic and should be handled with great care. Gloves and a face protector should be worn.

1000x Mineral stock solution:

$FeCl_3$ (2 mM)	32 mg
$MgSO_4 \cdot 7H_2O$ (0.8 M)	19.72 g
Distilled H_2O	100 ml

Autoclave and store. This solution will form a fine yellow precipitate, which should be resuspended before use.

100x Vitamin stock solution:

Thiamine (0.04 mg/ml)	4 mg
Biotin (2 μg/ml)	0.2 mg
Pyridoxine (0.04 mg/ml)	4 mg
Inositol (0.2 mg/ml)	20 mg
Pantothenic acid (0.04 mg/ml)	4 mg
Distilled H_2O	100 ml

Filter-sterilize using a 0.2-μm filter.

X-Gal Plates for Lysed Yeast Cells on Filters

These plates are used for checking β-galactosidase activity in cells that have been lysed and are immobilized on 3MM filters.

Bacto-agar	20 g
1 M Na_2HPO_4	57.7 ml
1 M NaH_2PO_4	42.3 ml
$MgSO_4$	0.25 g
Distilled H_2O	900 ml

After autoclaving, add 6 ml of X-Gal solution (20 mg/ml in *N,N*-dimethylformamide).

PRESPORULATION MEDIUM

Strains are grown for 1–2 days on this medium before transferring them to sporulation medium. This is only necessary for strains that do not sporulate well when put in sporulation medium directly.

Bacto-yeast extract (0.8%)	0.8 g
Bacto-peptone (0.3%)	0.3 g
Glucose (10%)	10 g
Bacto-agar (2%)	2 g
Distilled H_2O	100 ml

SPORULATION MEDIUM

Strains will undergo several divisions on this medium and then sporulate after 3–5 days of incubation.

Potassium acetate (1%)	10 g
Bacto-yeast extract (0.1%)	1 g
Glucose (0.05%)	0.5 g
Bacto-agar (2%)	20 g
Distilled H_2O	1000 ml

Nutritional supplements are required for sporulation of auxotrophic diploids on sporulation medium. Supplements at the level of 25% of those used for SMM plates should be added when compounding the medium. Alternatively, supplements can be spread on the surface of the sporulation plates in the volumes listed for SMM. The liquid should be allowed to dry thoroughly onto the agar before inoculating it with yeast strains.

MINIMAL SPORULATION MEDIUM

*MAT***a**/*MAT*α diploid cells will sporulate on this medium after 18–24 hours without vegetative growth.

Potassium acetate (1%)	10 g
Bacto-agar (2%)	20 g
Distilled H_2O	1000 ml

Nutritional supplements are required for sporulation of auxotrophic diploids on sporulation medium. Supplements at the level of 25% of those used for SMM plates should be added when compounding the medium. Alternatively, supplements can be spread on the surface of the sporulation plates in the volumes listed for SMM. The liquid should be allowed to dry thoroughly onto the agar before inoculating it with yeast strains.

LOW-pH BLUE PLATES

These plates are used for testing killer phenotype.

Bacto-yeast extract (1%)	6 g
Bacto-peptone (2%)	12 g
Glucose (2%)	12 g
Bacto-agar (2%)	12 g
Distilled H_2O	533 ml
Autoclave the above ingredients and add the following solutions:	
Methylene blue in sterile H_2O	5 ml
Phosphate-citrate buffer (sterile)	67 ml

Methylene blue in sterile H_2O:

Methylene blue	20 mg
Sterile H_2O	5 ml

Phosphate-citrate buffer for low-pH medium:

Citric acid	14.07 g
K_2HPO_4	18.96 g
Distilled H_2O	67 ml

Adjust the pH to 4.5 using solid K_2HPO_4 or citric acid. Sterilize by autoclaving.

DRUG SELECTION MEDIA

5-Fluoro-orotic Acid Medium

5-Fluoro-orotic acid (5-FOA) can be used to select for mutant cells that fail to utilize orotic acid as the source of the pyrimidine ring. Wild-type cells convert 5-FOA to 5-fluoro-orotidine monophosphate by conjugation to phosphoribosyl pyrophosphate (PRPP) and subsequently decarboxylate it to form 5-fluoro-uridine monophosphate (5-FUMP). These two steps are catalyzed by the products of the yeast genes *URA5* and *URA3*, respectively. Inevitably, fluorodeoxyuridine formed later is a potent inhibitor of thymidylate synthetase and thereby quite toxic to the cell. The two steps of de novo synthesis of uridine that are required to convert 5-FOA to 5-FUMP can be mutated to block utilization, as long as uracil is provided to allow formation of UMP via the salvage pathway. Therefore, both *ura3*$^-$ and *ura5*$^-$ mutants can grow on 5-FOA-containing medium (Boeke et al. *Mol. Gen. Genet.* **197:** 345 [1984]). In practice, only *ura3*$^-$ mutants appear to be uracil auxotrophs. The enzyme that catalyzes conjugation of uracil to PRPP can utilize orotic acid as a substrate at some level allowing *ura5*$^-$ mutants to grow slowly in the absence of uracil.

Bacto-yeast nitrogen base (0.67%)	6.7 g
Drop-out mix – ura (0.2%)	2 g
Glucose (2%)	20 g
Uracil (50 µg/ml)	50 mg
5-FOA (0.1%)	1 g
Distilled H_2O	500 ml

Dissolve the above and filter-sterilize using a 0.2-µm filter.

Autoclave the agar separately:

Bacto-agar (2%)	20 g
Distilled H_2O	500 ml

Mix the two solutions after cooling the agar to ~80°C. Pour into petri dishes (25 ml/dish).

5-FOA Medium ala HC

Make one liter of HC plates as described above, let flask cool to 50°C, then add 1.0 g 5-FOA.

α-Aminoadipate Plates

Wild-type strains are unable to utilize high levels of α-aminoadipate αAA as their sole nitrogen source because it is converted into a toxic intermediate by the normal lysine anabolic pathway (Chattoo et al. *Genetics* **93:** 51–65 [1979]; Zaret and Sherman *J. Bacteriol.* **162:** 579–583 [1985]). This medium is frequently used in the selection of mutations in the *LYS2* and *LYS5* genes.

Bacto-yeast nitrogen base without amino acids or ammonium sulfate (0.16%)	1.6 g
Glucose (2%)	20 g
Lysine (30 mg/liter)	30 mg
Bacto-agar (2%)	20 g
Distilled H_2O	960 ml

Autoclave and add 40 ml of a 5% αAA solution.

5% αAA:

α-Aminoadipic acid	2 g
Distilled H_2O	40 ml

Mix and adjust the pH to 6 with 10 N KOH to allow dissolution. Filter-sterilize using a 0.2-μm filter before adding to the autoclaved ingredients. **Safety Note:** Concentrated bases should be handled with great care; gloves and a face protector should be worn.

Cycloheximide

Cycloheximide resistance can arise in a number of different genes, but resistance to high levels ordinarily occurs due to rare mutations at the *cyh2* locus, which encodes the L29 ribosomal subunit. Resistance to cycloheximide is recessive, presumably because the sensitive ribosomes remain bound to the mRNA and block all further elongation.

Cycloheximide can be used in either YPD or synthetic media. A final concentration of 10 mg/liter should be used for YPD and 3 mg/liter for SD, SC, and YPG. A stock solution is prepared by dissolving 100 mg of cycloheximide in 10 ml of distilled H_2O and then filter-sterilizing (0.2-μm filter). The stock solution can be stored at 4°C. Appropriate volumes can be added to media after autoclaving.

Canavanine

Canavanine is an analog of arginine. Both are imported into the cell via the same high-affinity permease, which is encoded by the *can1* locus. High-level resistance to canavanine occurs exclusively because of mutation at this locus, but low-level resistance can arise at a number of other loci.

Because canavinine is a competitive inhibitor, arginine must be excluded from media used for testing sensitivity to the drug. Canavanine resistance must also be scored under high-nitrogen conditions, such as those provided by SD or SC medium, since the *CAN1* permease will then provide the only entry route to the cell for arginine and canavanine. In the presence of low-nitrogen conditions—effectively those provided by YPD medium—the general amino acid permease (*GAP*) system is induced and arginine and canavanine can also be taken up by this route. In addition, Can^R Arg^- auxotrophs are viable on YPD but are inviable on synthetic media because they are unable to take up arginine.

Canavanine sulfate is typically made up as a filter-sterilized (0.2-μm filter) 20 mg/ml stock solution in distilled H_2O. It is stored at 4°C and added to SD or SC – arg medium after autoclaving. A concentration of 60 mg/liter is typically used for scoring and selecting canavanine resistance.

Appendix B

Stock Preservation

Yeast strains can be stored indefinitely in 15% (v/v) glycerol at a temperature of −60°C or less. (Yeast tends to die if stored at temperatures above −55°C.) Many workers use 2-ml vials (35- x 12-mm) containing 1 ml of sterile 15% (v/v) glycerol. The strains are grown on the surfaces of YPD plates. The yeast is then scraped up with sterile applicator sticks or toothpicks and suspended in the glycerol solution. The caps are tightened and the vials shaken before freezing. The yeast can be revived by transferring a small portion of the frozen sample onto a YPD plate.

Yeast strains can be stored at 4°C for up to 6 months on slants prepared with YPAD medium. This method of storage is convenient since the slants take up little space, do not dry out, and contain excess adenine to prevent toxicity due to the red pigment produced by certain *ade*⁻ mutants. Slants are also a useful means of sending strains to colleagues.

REFERENCE

Well, A.M. and G.G. Stewart. 1973. Storage of brewing yeasts by liquid nitrogen refrigeration. *Appl. Microbiol.* **26:** 577.

Appendix C

Yeast Genetic and Physical Maps

Figures I–XVI on the following pages (162–167) depict the genetic and physical maps, and their correlations, of the 16 *Saccharomyces cerevisiae* chromosomes. A parallel comparison of the physical map (left, in kilobase pairs) and the genetic map (right, in centimorgans) of each of the 16 chromosomes is illustrated. The information in this figure is available on the *Saccharomyces* Genome Database (http://genome-www.stanford.edu/Saccharomyces/). The physical map consists of shaded boxes that indicate ORFs. ORFs on the Watson strand (left telomere is the 5′ end of this strand) are shown as light-gray boxes; those on the Crick strand as dark-gray boxes. Where it has been defined, the gene name of an ORF is indicated. The genetic map is based on data collected since 1991 by the SGD project as well as earlier data. Horizonal tick marks on the right of the genetic map line indicate positions of genes. Lines connect genetically mapped genes with their ORF on the physical map. A single name is listed for known synonyms. (Reprinted, with permission, from Cherry et al. [1997] *Nature* **387:** [supplement] 67–73. Copyright Macmillan Magazines Ltd.)

REFERENCE

Cherry, J.M., C. Ball, S. Weng, G. Juvik, R. Schmidt, C. Adler, B. Dunn, S. Dwight, L. Riles, R.K. Mortimer, and D. Botstein. 1997. Genetic and physical maps of *Saccharomyces cerevisiae*. *Nature* **387:** (suppl.) 67–73.

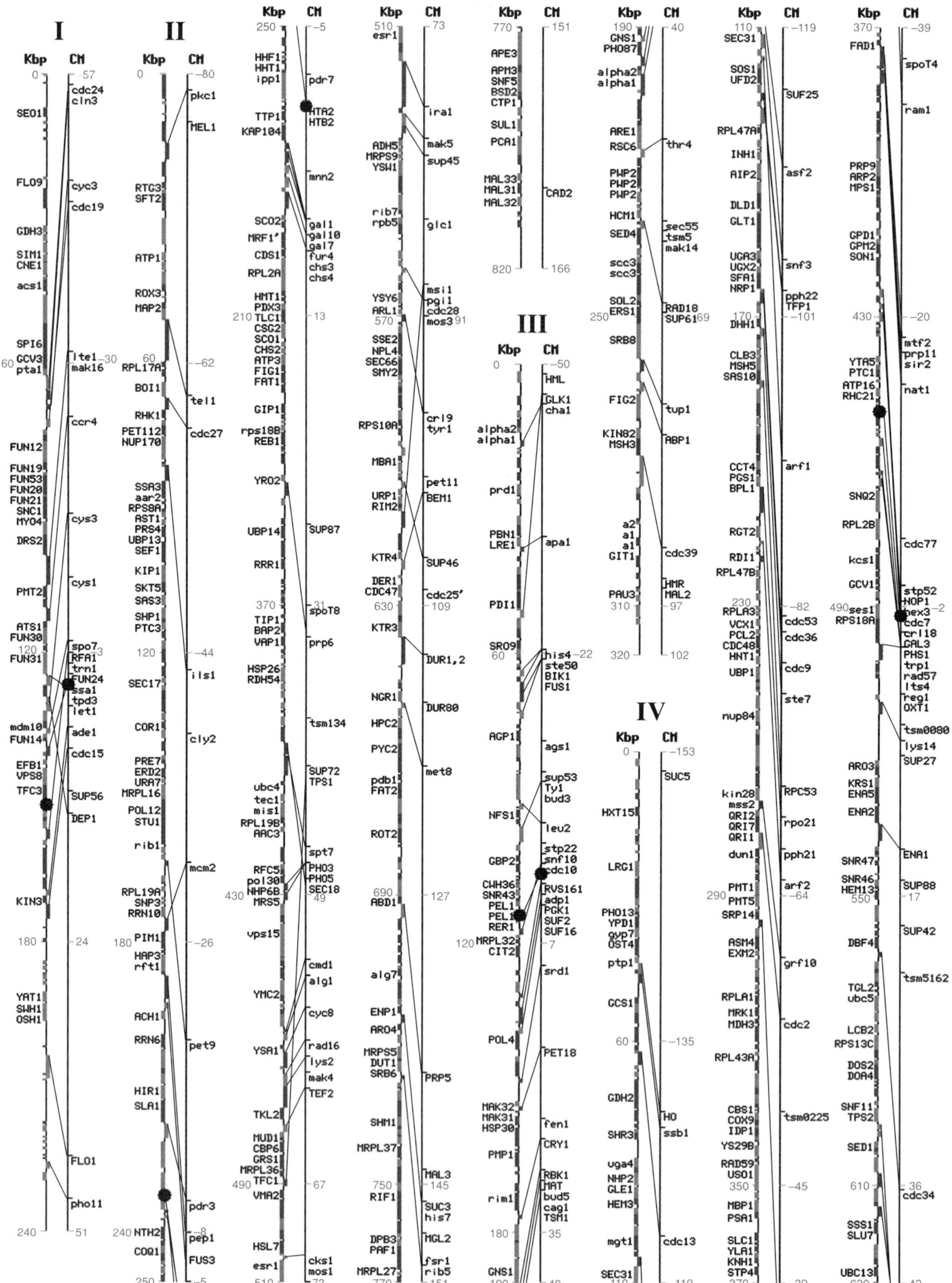

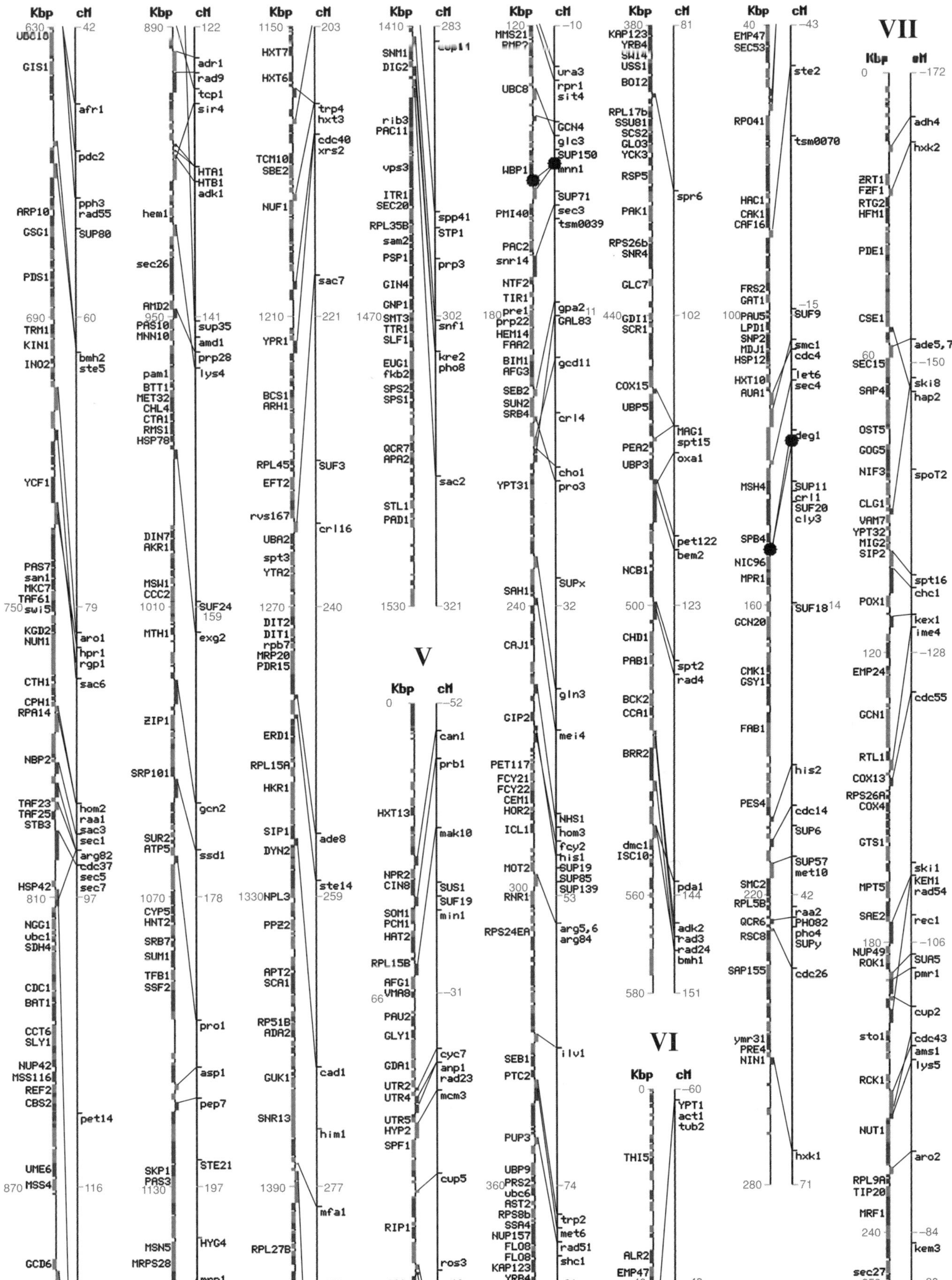

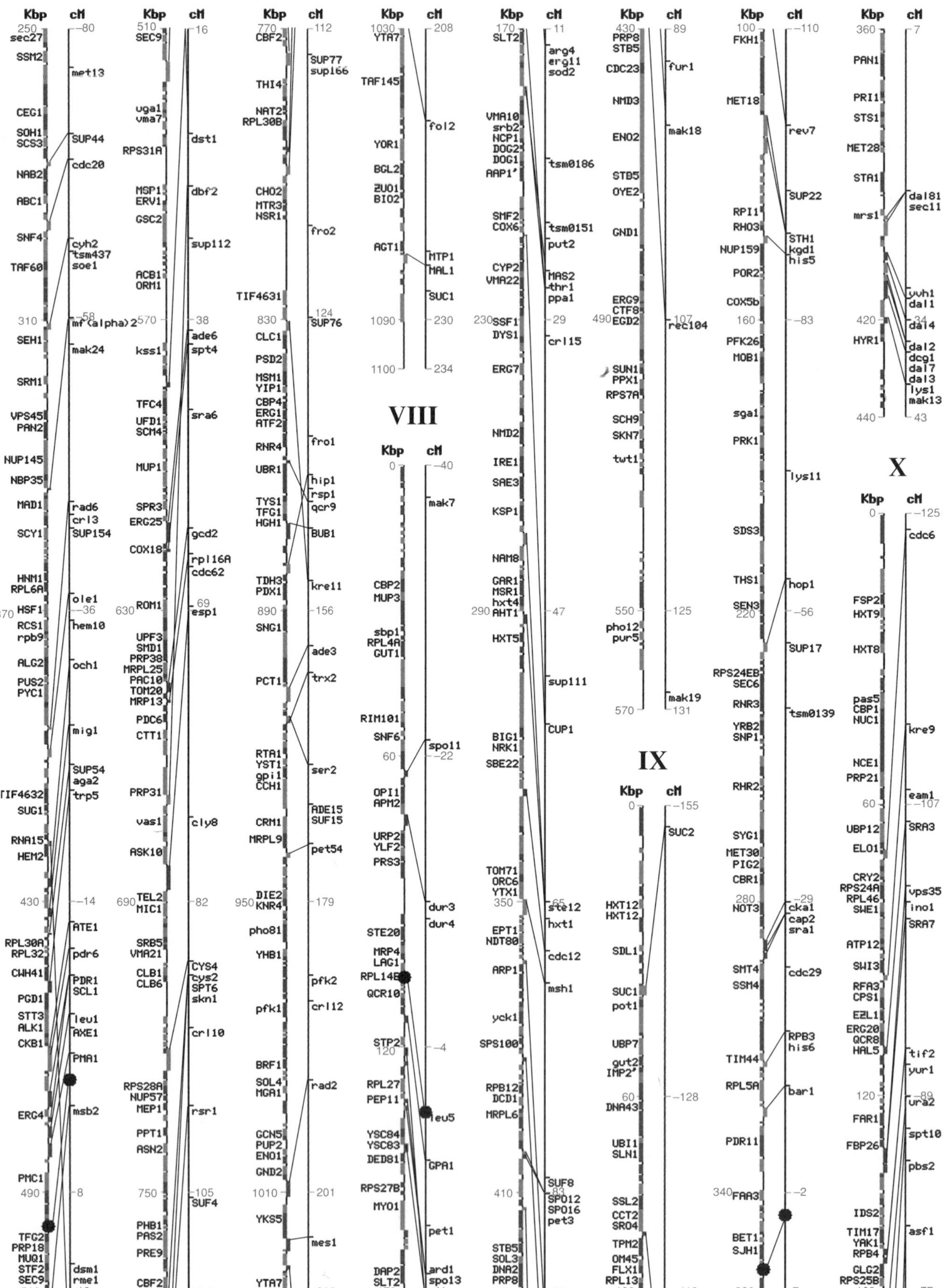

Kbp
cM
250
sec27
SSM2
CEG1
SOH1
SCS3
NAB2
ABC1
SNF4
TAF60
310
SEH1
SRM1
VPS45
PAN2
NUP145
NBP35
MAD1
SCY1
HNM1
RPL6A
370
HSF1
RCS1
rpb9
ALG2
PUS2
PYC1
TIF4632
SUG1
RNA15
HEM2
430
RPL30A
RPL32
CWH41
PGD1
STT3
ALK1
CKB1
ERG4
PMC1
490
TFG2
PRP18
MUQ1
STF2
SEC9
510
80
met13
SUP44
cdc20
cyh2
tsm437
soe1
58
mf(alpha)2
mak24
rad6
crl3
SUP154
ole1
36
hem10
och1
mig1
SUP54
aga2
trp5
14
ATE1
pdr6
PDR1
SCL1
leu1
AXE1
PMA1
msb2
8
dsm1
rme1
16
Kbp
cM
510
SEC9
uga1
vma7
RPS31A
MSP1
ERV1
GSC2
ACB1
ORM1
570
kss1
TFC4
UFD1
SCM4
MUP1
SPR3
ERG25
COX18
ROM1
630
UPF3
SMD1
PRP38
MRPL25
PAC10
TOM20
MRP13
PDC6
CTT1
PRP31
vas1
ASK10
TEL2
690
MIC1
SRB5
VMA21
CLB1
CLB6
RPS28A
NUP57
MEP1
PPT1
ASN2
750
PHB1
PAS2
PRE9
CBF2
770
16
dst1
dbf2
sup112
38
ade6
spt4
sra6
gcd2
rpl16A
cdc62
69
esp1
cly8
82
CYS4
cys2
SPT6
skn1
crl10
rsr1
105
SUF4
112
Kbp
cM
770
CBF2
THI4
NAT2
RPL30B
CHO2
MTR3
NSR1
TIF4631
830
CLC1
PSD2
MSM1
YIP1
CBP4
ERG1
ATF2
RNR4
UBR1
TYS1
TFG1
HGH1
TDH3
PDX1
890
SNG1
PCT1
RTA1
YST1
gpi1
CCH1
CRM1
MRPL9
DIE2
950
KNR4
pho81
YHB1
pfk1
BRF1
SOL4
MGA1
GCN5
PUP2
ENO1
GND2
1010
YKS5
YTA7
1030
112
SUP77
sup166
fro2
124
SUP76
fro1
hip1
rsp1
qcr9
BUB1
kre11
156
ade3
trx2
ser2
ADE15
SUF15
pet54
179
pfk2
crl12
rad2
201
mes1
208
Kbp
cM
1030
YTA7
TAF145
YOR1
BGL2
ZUO1
BIO2
AGT1
1090
1100
208
fol2
MTP1
MAL1
SUC1
230
234
VIII
Kbp
cM
0
CBP2
MUP3
sbp1
RPL4A
GUT1
RIM101
SNF6
60
OPI1
APM2
URP2
YLF2
PRS3
STE20
MRP4
LAG1
RPL14B
QCR10
STP2
120
RPL27
PEP11
YSC84
YSC83
DED81
RPS27B
MYO1
DAP2
SLT2
170
40
mak7
spo11
22
dur3
dur4
4
leu5
GPA1
pet1
ard1
spo13
11
Kbp
cM
170
SLT2
VMA10
srb2
NCP1
DOG2
DOG1
AAP1'
SMF2
COX6
CYP2
VMA22
230
SSF1
DYS1
ERG7
NMD2
IRE1
SAE3
KSP1
NAM8
GAR1
MSR1
hxt4
290
AHT1
HXT5
BIG1
NRK1
SBE22
TOM71
ORC6
YTX1
350
EPT1
NDT80
ARP1
yck1
SPS100
RPB12
DCD1
MRPL6
410
STB5
SOL3
DNA2
PRP8
430
11
arg4
erg11
sod2
tsm0186
tsm0151
put2
MAS2
thr1
ppa1
29
crl15
47
sup111
CUP1
65
ste12
hxt1
cdc12
msh1
83
SUF8
SPO12
SPO16
pet3
89
Kbp
cM
430
PRP8
STB5
CDC23
NMD3
ENO2
STB5
OYE2
GND1
ERG9
CTF8
490
EGD2
SUN1
PPX1
RPS7A
SCH9
SKN7
twt1
550
pho12
pur5
570
89
fur1
mak18
107
rec104
125
mak19
131
IX
Kbp
cM
0
HXT12
HXT12
SDL1
SUC1
pot1
UBP7
gut2
IMP2'
60
DNA43
UBI1
SLN1
SSL2
CCT2
SRO4
TPM2
OM45
FLX1
RPL13
100
155
SUC2
128
110
Kbp
cM
100
FKH1
MET18
RPI1
RHO3
NUP159
POR2
COX5b
160
PFK26
MOB1
sga1
PRK1
SDS3
THS1
SEN3
220
RPS24EB
SEC6
RNR3
YRB2
SNP1
RHR2
SYG1
MET30
PIG2
CBR1
280
NOT3
SMT4
SSM4
TIM44
RPL5A
PDR11
340
FAA3
BET1
SJH1
360
110
rev7
SUP22
STH1
kgd1
his5
83
lys11
hop1
56
SUP17
tsm0139
29
cka1
cap2
sra1
cdc29
RPB3
his6
bar1
2
7
Kbp
cM
360
PAN1
PRI1
STS1
MET28
STA1
mrs1
420
HYR1
440
7
dal81
sec11
yvh1
dal1
34
dal4
dal2
dcg1
dal7
dal3
lys1
mak13
43
X
Kbp
cM
0
FSP2
HXT9
HXT8
pas5
CBP1
NUC1
NCE1
PRP21
60
UBP12
ELO1
CRY2
RPS24A
RPL46
SWE1
ATP12
SWI3
RFA3
CPS1
EZL1
ERG20
QCR8
HAL5
120
FAR1
FBP26
IDS2
TIM17
YAK1
RPB4
GLG2
RPS25B
160
125
cdc6
kre9
eam1
107
SRA3
vps35
ino1
SRA7
tif2
yur1
89
ura2
spt10
pbs2
asf1
77

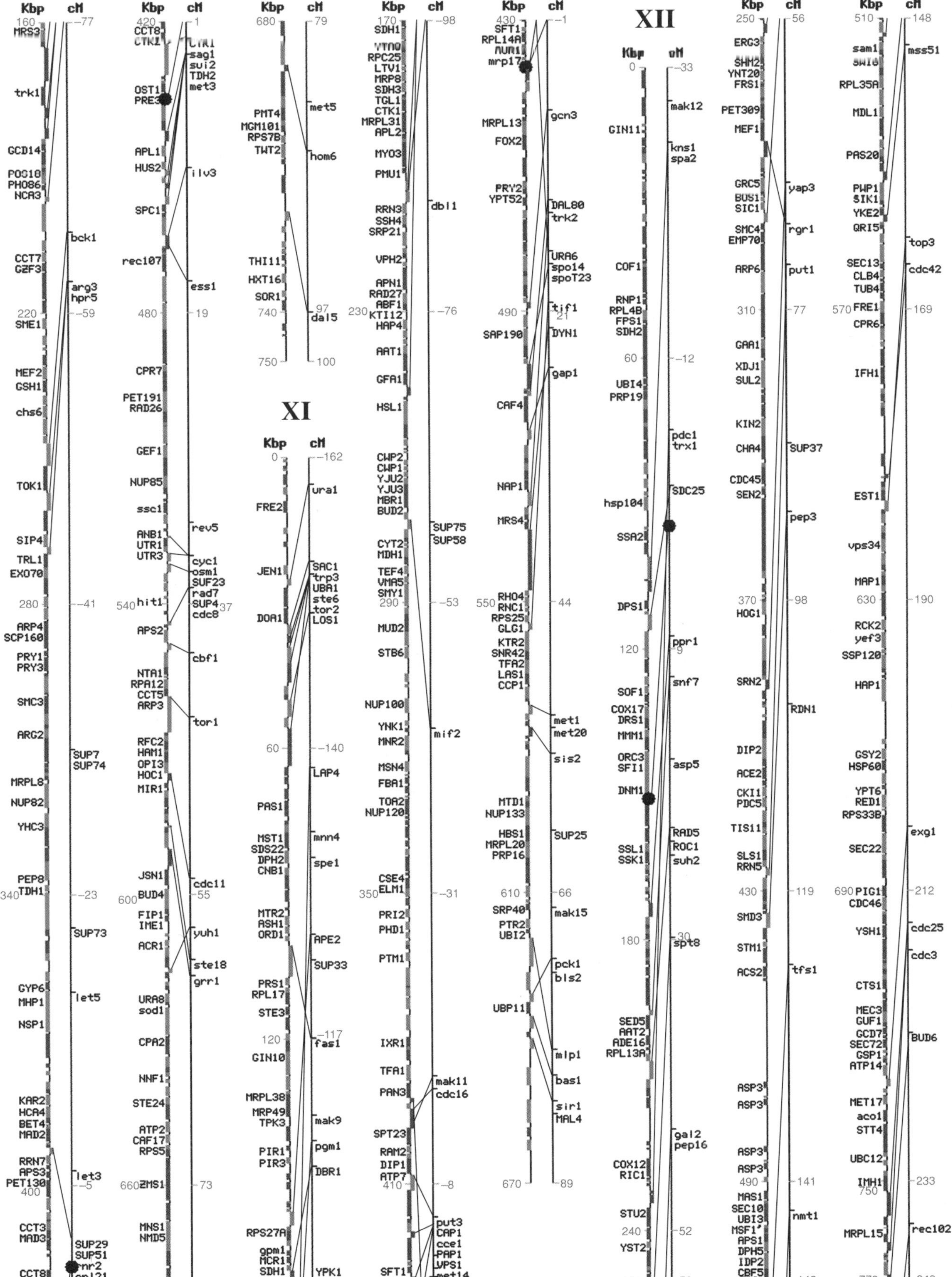

Kbp cM
160 77
MRS3
trk1
GCD14
POS18
PHO86
NCA3
bck1
CCT7
GZF3
arg3
hpr5
220 59
SME1
MEF2
GSH1
chs6
TOK1
SIP4
TRL1
EXO70
280 41
ARP4
SCP160
PRY1
PRY3
SMC3
ARG2
SUP7
SUP74
MRPL8
NUP82
YHC3
PEP8
TDH1
340 23
SUP73
GYP6
MHP1
let5
NSP1
KAR2
HCA4
BET4
MAD2
RRN7
APS3
PET130
let3
400 5
CCT3
MAD3
SUP29
SUP51
rnr2
crl21
CCT8
420 1
Kbp cM
420 1
CCT8
CTK2
sag1
sui2
TDH2
met3
OST1
PRE3
APL1
HUS2
ilv3
SPC1
rec107
ess1
480 19
CPR7
PET191
RAD26
GEF1
NUP85
ssc1
rev5
ANB1
UTR1
UTR3
cyc1
osm1
SUF23
rad7
540 hit1 37
SUP4
cdc8
APS2
cbf1
NTA1
RPA12
CCT5
ARP3
tor1
RFC2
HAM1
OPI3
HOC1
MIR1
JSN1
cdc11
600 BUD4 55
FIP1
IME1
yuh1
ACR1
ste18
grr1
URA8
sod1
CPA2
NNF1
STE24
ATP2
CAF17
RPS5
660 ZMS1 73
MNS1
NMD5
680 79
Kbp cM
680 79
met5
PMT4
MGM101
RPS7B
TWT2
hom6
THI11
HXT16
SOR1
740 97
dal5
750 100
XI
Kbp cM
0 162
ura1
FRE2
JEN1
SAC1
trp3
UBA1
ste6
tor2
LOS1
DOA1
60 140
LAP4
PAS1
MST1
mnn4
SDS22
DPH2
spe1
CNB1
MTR2
ASH1
ORD1
APE2
SUP33
PRS1
RPL17
STE3
120 117
fas1
GIN10
MRPL38
MRP49
TPK3
mak9
PIR1
pgm1
PIR3
DBR1
RPS27A
gpm1
MCR1
SDH1
YPK1
170 98
Kbp cM
170 98
SDH1
RPC25
LTV1
MRP8
SDH3
TGL1
CTK1
MRPL31
APL2
MYO3
PMU1
dbl1
RRN3
SSH4
SRP21
VPH2
APN1
RAD27
ABF1
230 KTI12 76
HAP4
AAT1
GFA1
HSL1
CWP2
CWP1
YJU2
YJU3
MBR1
BUD2
SUP75
SUP58
CYT2
MDH1
TEF4
VMA5
SMY1
290 53
MUD2
STB6
NUP100
YNK1
MNR2
mif2
MSN4
FBA1
TOA2
NUP120
CSE4
350 ELM1 31
PRI2
PHD1
PTM1
IXR1
TFA1
mak11
PAN3
cdc16
SPT23
RAM2
DIP1
ATP7
410 8
put3
CAP1
cce1
PAP1
VPS1
SFT1
met14
430 1
Kbp cM
430 1
SFT1
RPL14A
mrp17
gcn3
MRPL13
FOX2
PRY2
YPT52
DAL80
trk2
URA6
spo14
spoT23
tif1
490 21
SAP190
DYN1
gap1
CAF4
NAP1
MRS4
RHO4
550 RNC1 44
RPS25
GLG1
KTR2
SNR42
TFA2
LAS1
CCP1
met1
met20
sis2
MTD1
NUP133
HBS1
SUP25
MRPL20
PRP16
610 66
SRP40
mak15
PTR2
UBI2
pck1
bls2
UBP11
mlp1
bas1
sir1
MAL4
670 89
XII
Kbp cM
0 33
mak12
GIN11
kns1
spa2
COF1
RNP1
RPL4B
FPS1
SDH2
60 12
UBI4
PRP19
pdc1
trx1
hsp104
SDC25
SSA2
DPS1
120 9
ppr1
snf7
SOF1
COX17
DRS1
MMM1
ORC3
SFI1
asp5
DNM1
RAD5
SSL1
ROC1
SSK1
suh2
180 30
spt8
SED5
AAT2
ADE16
RPL13A
gal2
pep16
COX12
RIC1
STU2
240 52
YST2
250 56
Kbp cM
250 56
ERG3
YNT20
FRS1
PET309
MEF1
GRC5
yap3
SIC1
rgr1
SMC4
EMP70
ARP6
put1
310 77
GAA1
XDJ1
SUL2
KIN2
CHA4
SUP37
CDC45
SEN2
pep3
370 98
HOG1
SRN2
RDN1
DIP2
ACE2
CKI1
PDC5
TIS11
SLS1
RRN5
430 119
SMD3
STM1
ACS2
tfs1
ASP3
ASP3
ASP3
ASP3
490 141
MAS1
SEC10
nmt1
UBI3
MSF1
APS1
DPH5
IDP2
CBF5
510 148
Kbp cM
510 148
sam1
mss51
RPL35A
MDL1
PAS20
PWP1
SIK1
YKE2
QRI5
top3
SEC13
CLB4
cdc42
TUB4
570 FRE1 169
CPR6
IFH1
EST1
vps34
MAP1
630 190
RCK2
yef3
SSP120
HAP1
GSY2
HSP60
YPT6
RED1
RPS33B
exg1
SEC22
690 PIG1 212
CDC46
YSH1
cdc25
cdc3
CTS1
MEC3
GUF1
GCD7
BUD6
SEC72
GSP1
ATP14
MET17
aco1
STT4
UBC12
IMH1 233
750
MRPL15
rec102
770 240

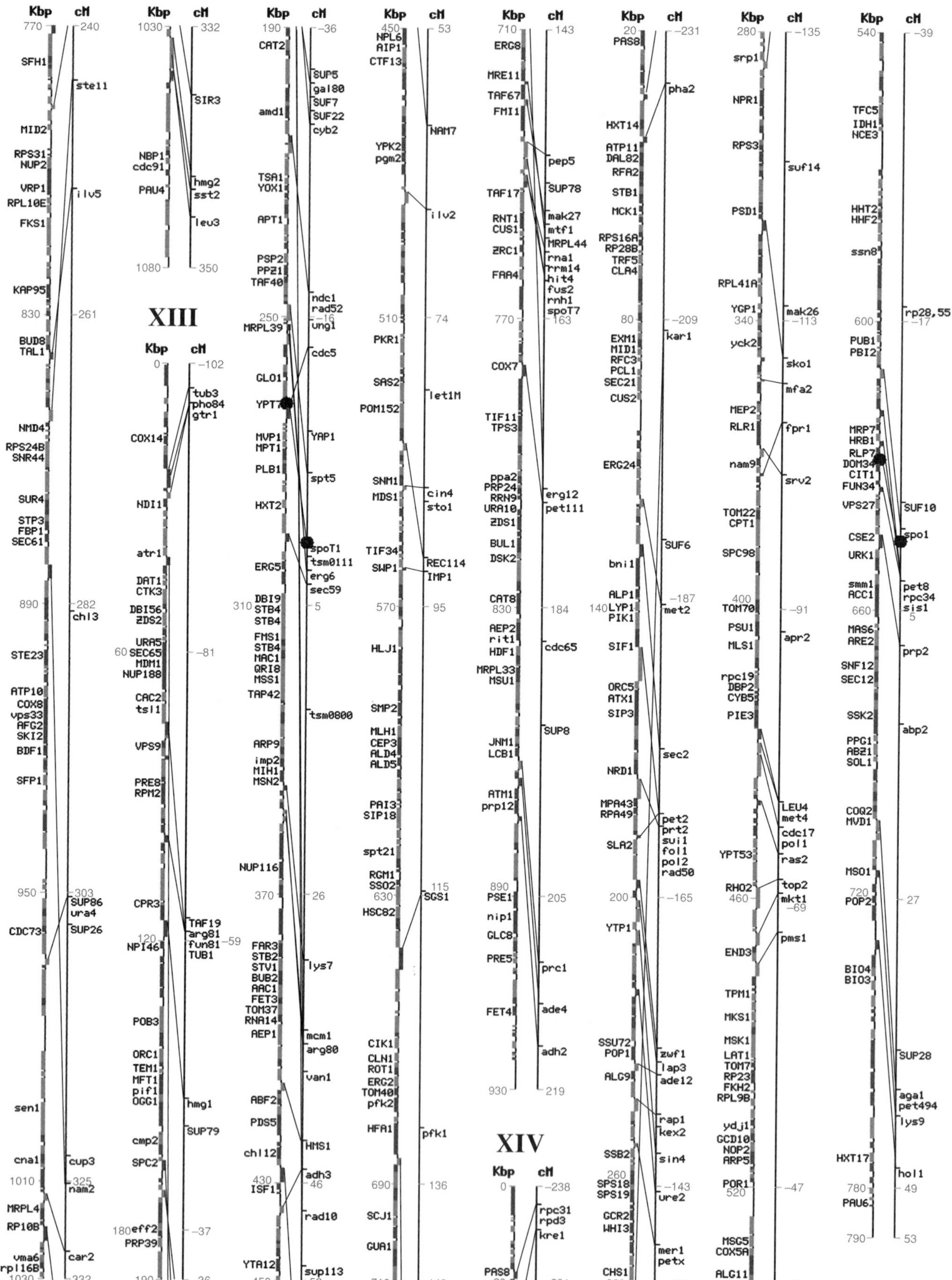
Kbp cM
770 240 SFH1 ste11 MID2 RPS31 NUP2 VRP1 RPL10E ilv5 FKS1 KAP95 830 261 BUD8 TAL1 NMD4 RPS24B SNR44 SUR4 STP3 FBP1 SEC61 890 282 chl3 STE23 ATP10 COX8 vps33 AFG2 SKI2 BDF1 SFP1 950 303 SUP86 ura4 CDC73 SUP26 sen1 cna1 cup3 1010 325 nam2 MRPL4 RP10B car2 vma6 rpl16B 1030 332
Kbp cM
1030 332 SIR3 NBP1 cdc91 PAU4 hmg2 sst2 leu3 1080 350
XIII
Kbp cM
0 -102 tub3 pho84 gtr1 COX14 NDI1 atr1 DAT1 CTK3 DBI56 ZDS2 URA5 60 SEC65 -81 MDM1 NUP188 CAC2 tsl1 VPS9 PRE8 RPM2 CPR3 TAF19 arg81 120 NPI46 fun81 -59 TUB1 POB3 ORC1 TEM1 MFT1 pif1 OGG1 hmg1 SUP79 cmp2 SPC2 180 eff2 -37 PRP39 190 -36
Kbp cM
190 -36 CAT2 SUP5 gal80 SUF7 SUF22 amd1 cyb2 TSA1 YOX1 APT1 PSP2 PPZ1 TAF40 ndc1 rad52 250 -16 ung1 MRPL39 cdc5 GLO1 YPT7 YAP1 MVP1 MPT1 PLB1 spt5 HXT2 spoT1 tsm0111 ERG5 erg6 sec59 DBI9 310 STB4 5 STB4 FMS1 STB4 MAC1 QRI8 MSS1 TAP42 tsm0800 ARP9 imp2 MIH1 MSN2 NUP116 370 26 FAR3 STB2 STV1 BUB2 AAC1 FET3 TOM37 RNA14 AEP1 lys7 mcm1 arg80 van1 ABF2 PDS5 HMS1 chl12 430 adh3 ISF1 46 rad10 YTA12 sup113 450 53
Kbp cM
450 53 NPL6 AIP1 CTF13 NAM7 YPK2 pgm2 ilv2 510 74 PKR1 SAS2 let1M POM152 SNM1 cin4 MDS1 sto1 TIF34 REC114 SWP1 IMP1 570 95 HLJ1 SMP2 MLH1 CEP3 ALD4 ALD5 PAI3 SIP18 spt21 RGM1 SSO2 115 630 SGS1 HSC82 CIK1 CLN1 ROT1 ERG2 TOM40 pfk2 HFA1 pfk1 690 136 SCJ1 GUA1 710 143
Kbp cM
710 143 ERG8 MRE11 TAF67 FMI1 pep5 SUP78 TAF17 mak27 RNT1 mtf1 CUS1 MRPL44 ZRC1 rna1 rrm14 FAA4 hit4 fus2 rnh1 spoT7 770 163 COX7 TIF11 TPS3 ppa2 PRP24 erg12 RRN9 pet111 URA10 ZDS1 BUL1 DSK2 CAT8 830 184 AEP2 rit1 cdc65 HDF1 MRPL33 MSU1 SUP8 JNM1 LCB1 ATM1 prp12 890 PSE1 205 nip1 GLC8 PRE5 prc1 FET4 ade4 adh2 930 219
XIV
Kbp cM
0 -238 rpc31 rpd3 kre1 PAS8 20 -231
Kbp cM
20 -231 PAS8 pha2 HXT14 ATP11 DAL82 RFA2 STB1 MCK1 RPS16A RP28B TRF5 CLA4 80 -209 kar1 EXM1 MID1 RFC3 PCL1 SEC21 CUS2 ERG24 SUF6 bni1 ALP1 -187 140 LYP1 met2 PIK1 SIF1 ORC5 ATX1 SIP3 sec2 NRD1 MPA43 RPA49 pet2 prt2 sui1 SLA2 fol1 pol2 rad50 200 -165 YTP1 SSU72 zwf1 POP1 lap3 ALG9 ade12 rap1 kex2 SSB2 sin4 260 SPS18 -143 SPS19 ure2 GCR2 WHI3 mer1 petx CHS1 280 -135
Kbp cM
280 -135 srp1 NPR1 RPS3 suf14 PSD1 RPL41A YGP1 mak26 340 -113 yck2 sko1 mfa2 MEP2 RLR1 fpr1 nam9 srv2 TOM22 CPT1 SPC98 400 TOM70 -91 PSU1 MLS1 apr2 rpc19 DBP2 CYB5 PIE3 LEU4 met4 cdc17 pol1 YPT53 ras2 RHO2 top2 460 mkt1 -69 pms1 END3 TPM1 MKS1 MSK1 LAT1 TOM7 RP23 FKH2 RPL9B ydj1 GCD10 NOP2 ARP5 POR1 520 -47 MSG5 COX5A ALG11 540 -39
Kbp cM
540 -39 TFC5 IDH1 NCE3 HHT2 HHF2 ssn8 600 rp28,55 -17 PUB1 PBI2 MRP7 HRB1 RLP7 DOM34 CIT1 FUN34 VPS27 SUF10 CSE2 spo1 URK1 smm1 pet8 ACC1 rpc34 660 sis1 5 MAS6 ARE2 prp2 SNF12 SEC12 SSK2 abp2 PPG1 ABZ1 SOL1 COQ2 MVD1 MSO1 720 27 POP2 BIO4 BIO3 SUP28 aga1 pet494 lys9 HXT17 hol1 780 49 PAU6 790 53

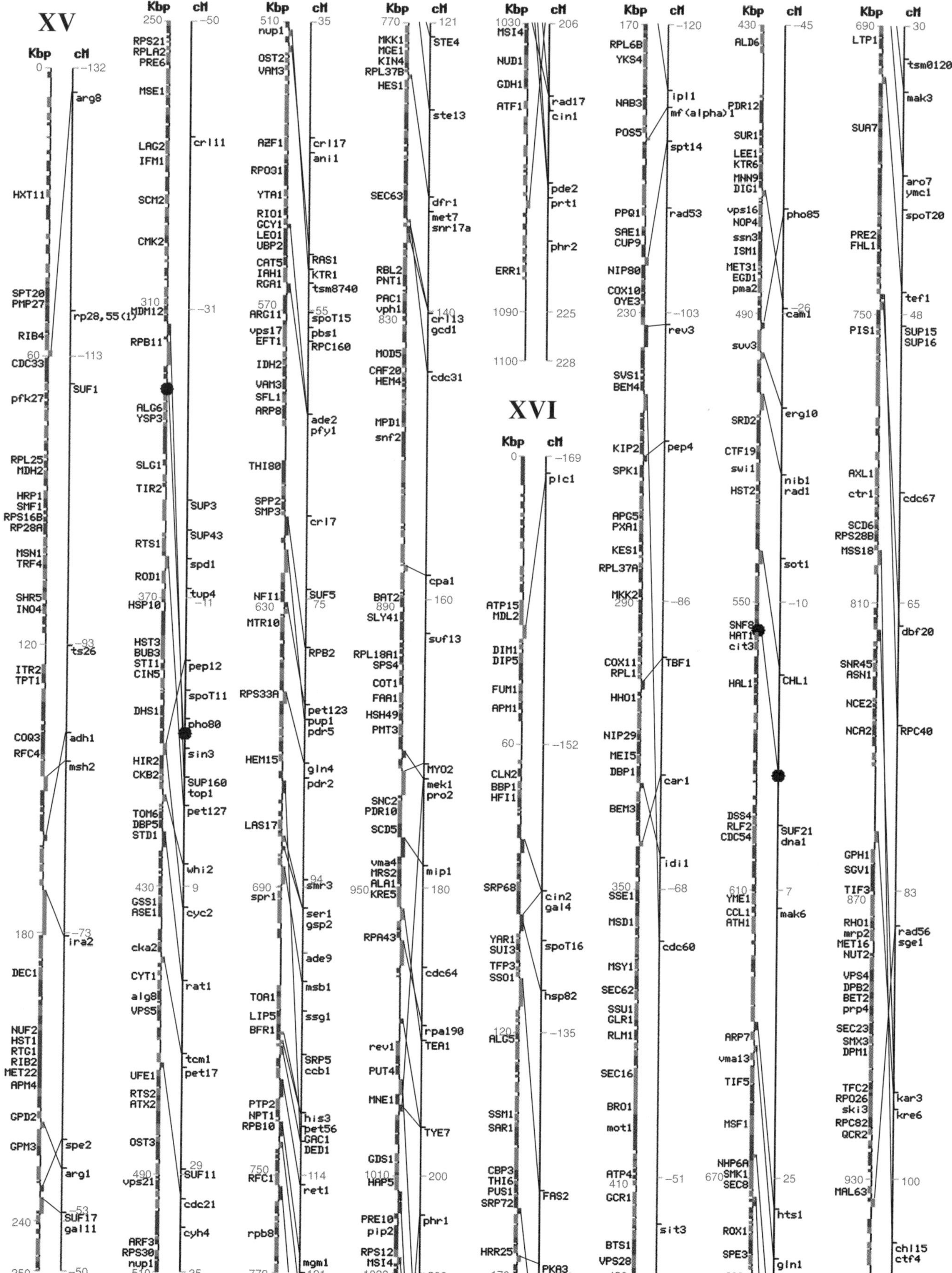

XV
Kbp
cM
0
-132
arg8
HXT11
SPT20
PMP27
rp28,55(1?)
RIB4
60
-113
CDC33
SUF1
pfk27
RPL25
MDH2
HRP1
SMF1
RPS16B
RP28A
MSN1
TRF4
SHR5
INO4
120
-93
ts26
ITR2
TPT1
COQ3
adh1
RFC4
msh2
180
-73
ira2
DEC1
NUF2
HST1
RTG1
RIB2
MET22
APM4
GPD2
GPM3
spe2
arg1
240
-53
SUF17
gal11
250
-50
Kbp
cM
250
-50
RPS21
RPLA2
PRE6
MSE1
LAG2
crl11
IFM1
SCM2
CMK2
310
MDM12
-31
RPB11
ALG6
YSP3
SLG1
TIR2
SUP3
SUP43
RTS1
spd1
ROD1
tup4
370
HSP10
-11
HST3
BUB3
STI1
CIN5
pep12
spoT11
DHS1
pho80
sin3
HIR2
CKB2
SUP160
top1
TOM6
DBP5
STD1
pet127
whi2
430
9
GSS1
ASE1
cyc2
cka2
CYT1
rat1
alg8
VPS5
tcm1
UFE1
pet17
RTS2
ATX2
OST3
29
490
vps21
SUF11
cdc21
cyh4
ARF3
RPS30
nup1
510
35
Kbp
cM
510
35
nup1
OST2
VAM3
AZF1
crl17
ani1
RPO31
YTA1
RIO1
GCY1
LEO1
UBP2
RAS1
CAT5
IAH1
KTR1
RGA1
tsm8740
570
55
ARG11
spoT15
vps17
pbs1
EFT1
RPC160
IDH2
VAM3
SFL1
ARP8
ade2
pfy1
THI80
SPP2
SMP3
crl7
NFI1
SUF5
630
75
MTR10
RPB2
RPS33A
pet123
pup1
pdr5
HEM15
gln4
pdr2
LAS17
690
94
smr3
spr1
ser1
gsp2
ade9
msb1
TOA1
LIP5
ssg1
BFR1
SRP5
ccb1
PTP2
NPT1
his3
RPB10
pet56
GAC1
DED1
750
114
RFC1
ret1
rpb8
mgm1
770
121
Kbp
cM
770
121
MKK1
STE4
MGE1
KIN4
RPL37B
HES1
ste13
SEC63
dfr1
met7
snr17a
RBL2
PNT1
PAC1
vph1
140
830
crl13
gcd1
MOD5
CAF20
HEM4
cdc31
MPD1
snf2
cpa1
BAT2
890
160
SLY41
suf13
RPL18A1
SPS4
COT1
FAA1
HSH49
PMT3
MYO2
mek1
pro2
SNC2
PDR10
SCD5
mip1
vma4
MRS2
ALA1
950
KRE5
180
RPA43
cdc64
rpa190
TEA1
rev1
PUT4
MNE1
TYE7
GDS1
1010
200
HAP5
PRE10
phr1
pip2
RPS12
MSI4
1030
206
Kbp
cM
1030
206
MSI4
NUD1
GDH1
ATF1
rad17
cin1
pde2
prt1
phr2
ERR1
1090
225
1100
228
XVI
Kbp
cM
0
-169
plc1
ATP15
MDL2
DIM1
DIP5
FUM1
APM1
60
-152
CLN2
BBP1
HFI1
SRP68
cin2
gal4
YAR1
SUI3
spoT16
TFP3
SSO1
hsp82
120
-135
ALG5
SSM1
SAR1
CBP3
THI6
PUS1
FAS2
SRP72
HRR25
PKA3
170
-120
Kbp
cM
170
-120
RPL6B
YKS4
ipl1
NAB3
mf(alpha)1
POS5
spt14
PPQ1
rad53
SAE1
CUP9
NIP80
COX10
OYE3
230
-103
rev3
SVS1
BEM4
KIP2
pep4
SPK1
APG5
PXA1
KES1
RPL37A
MKK2
290
-86
COX11
TBF1
RPL1
HHO1
NIP29
MEI5
DBP1
car1
BEM3
idi1
350
-68
SSE1
MSD1
cdc60
MSY1
SEC62
SSU1
GLR1
RLM1
SEC16
BRO1
mot1
ATP4
410
-51
GCR1
sit3
BTS1
VPS28
430
-45
Kbp
cM
430
-45
ALD6
PDR12
SUR1
LEE1
KTR6
MNN9
DIG1
vps16
pho85
NOP4
ssn3
ISM1
MET31
EGD1
pma2
-26
490
cam1
suv3
erg10
SRD2
CTF19
swi1
nib1
HST2
rad1
sot1
550
-10
SNF8
HAT1
cit3
HAL1
CHL1
DSS4
RLF2
SUF21
CDC54
dna1
610
7
YME1
CCL1
mak6
ATH1
ARP7
vma13
TIF5
MSF1
NHP6A
670
SMK1
25
SEC8
hts1
ROX1
SPE3
gln1
690
30
Kbp
cM
690
30
LTP1
tsm0120
mak3
SUA7
aro7
ymc1
spoT20
PRE2
FHL1
tef1
750
48
PIS1
SUP15
SUP16
AXL1
ctr1
cdc67
SCD6
RPS28B
MSS18
810
65
dbf20
SNR45
ASN1
NCE2
NCA2
RPC40
GPH1
SGV1
TIF3
870
83
RHO1
rad56
mrp2
sge1
MET16
NUT2
VPS4
DPB2
BET2
prp4
SEC23
SMX3
DPM1
TFC2
kar3
RPO26
ski3
kre6
RPC82
QCR2
930
100
MAL63
chl15
ctf4
950
106

APPENDIX D

Grids

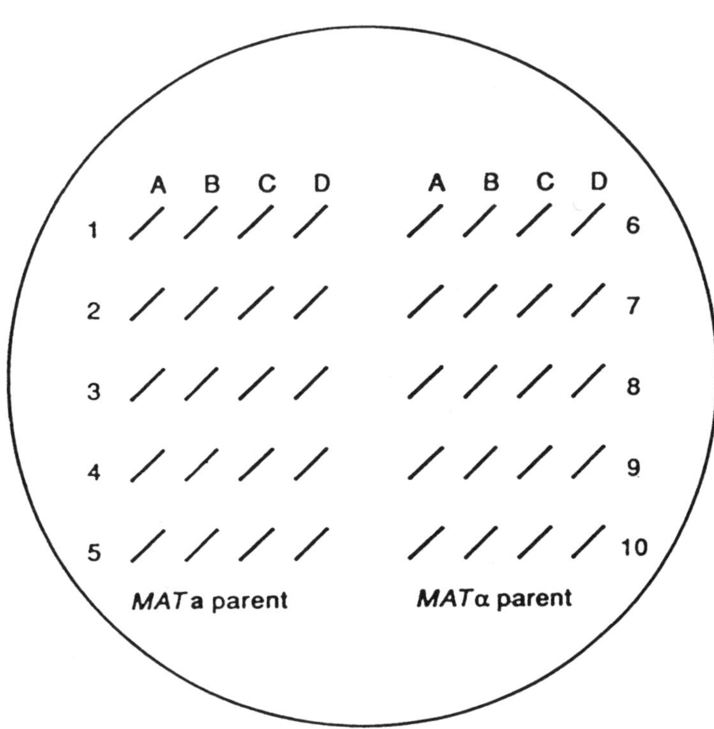

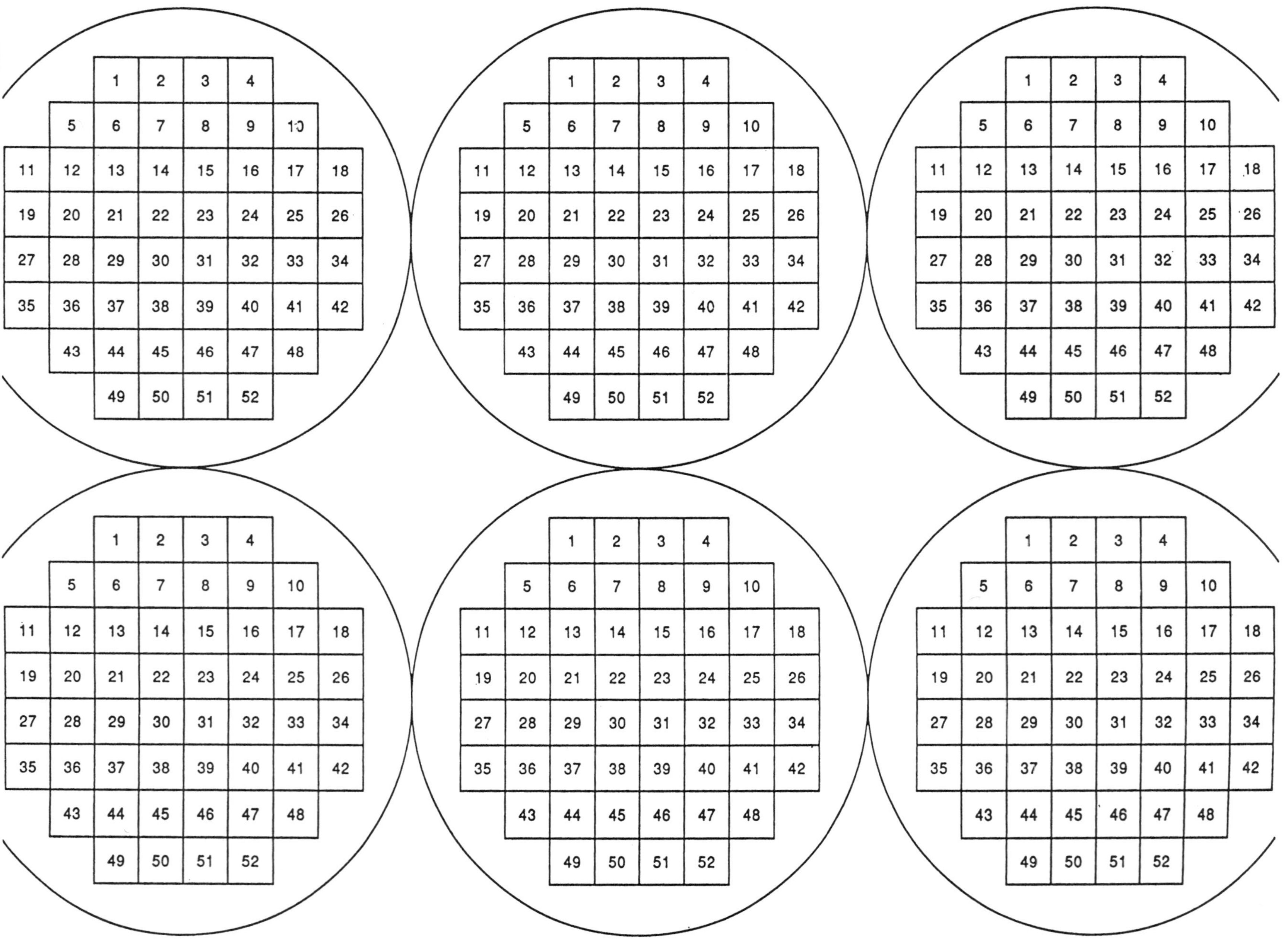
1 2 3 4
5 6 7 8 9 10
11 12 13 14 15 16 17 18
19 20 21 22 23 24 25 26
27 28 29 30 31 32 33 34
35 36 37 38 39 40 41 42
43 44 45 46 47 48
49 50 51 52
1 2 3 4
5 6 7 8 9 10
11 12 13 14 15 16 17 18
19 20 21 22 23 24 25 26
27 28 29 30 31 32 33 34
35 36 37 38 39 40 41 42
43 44 45 46 47 48
49 50 51 52
1 2 3 4
5 6 7 8 9 10
11 12 13 14 15 16 17 18
19 20 21 22 23 24 25 26
27 28 29 30 31 32 33 34
35 36 37 38 39 40 41 42
43 44 45 46 47 48
49 50 51 52
1 2 3 4
5 6 7 8 9 10
11 12 13 14 15 16 17 18
19 20 21 22 23 24 25 26
27 28 29 30 31 32 33 34
35 36 37 38 39 40 41 42
43 44 45 46 47 48
49 50 51 52
1 2 3 4
5 6 7 8 9 10
11 12 13 14 15 16 17 18
19 20 21 22 23 24 25 26
27 28 29 30 31 32 33 34
35 36 37 38 39 40 41 42
43 44 45 46 47 48
49 50 51 52
1 2 3 4
5 6 7 8 9 10
11 12 13 14 15 16 17 18
19 20 21 22 23 24 25 26
27 28 29 30 31 32 33 34
35 36 37 38 39 40 41 42
43 44 45 46 47 48
49 50 51 52

APPENDIX E

Electrophoretic Karyotypes of Strains for Southern Blot Mapping

Note: This appendix has been largely superseded by the sequencing of the yeast genome (see Appendix C), but this information is still useful for evaluating the size of whole chromosomes by pulsed-field gel electrophoresis.

CHROMOSOME ASSIGNMENTS

(From Carle and Olson [1985] *Proc. Natl. Acad. Sci.* **82:** 3756–3760.)

Indexed by Band (in order of decreasing electrophoretic mobility)

Band	Chromosome	Specific identifying hybridization probe
1	I	CDC19
2	VI	SUP11
3	III	SUP61
4	IX	SUP17
5A	VIII	ARG4
5B	V	URA3
6	XI	URA1
7	X	URA2
8	XIV	SUF10
9	II	LYS2
10A	XIII	SUP8
10B	XVI	GAL4
11A,B	XV,VII	SUP3, LEU1
12	IV	SUP2

Defining Strains

Strain	Resolved doublet	Genotype
AB972	(Standard)	*MATα trp1*
A364a	5A, 5B	*MAT***a** *ura1 lys2 ade2 ade1 his7 tyr1*
YPH45	10A, 10B	*MAT***a** *ura3-52 lys2 ade2 trp1-Δ1*

Mapping Strains

Strain	Genotype
YPH80	*MATα ura3-52 lys2 ade⁻ his7 trp1-Δ1*
YPH81	*MAT***a** *ura3-52 lys2 ade⁻ trp1-Δ1*

YPH149 and YPH152 are derived from YPH80 and YPH81, respectively, via two chromosome-fragmentation events. They have the following differences from the standard karyotype:

Band 11 equals chromosome XV only (no VII).

A chromosome fragment ($URA3^+$) derived from chromosome VII and carrying sequences centromere-proximal to RAD2 migrates between bands 10b and 11.

A chromosome fragment ($TRP1^+$) derived from chromosome VII and carrying sequences centromere-distal to RAD2 migrates below band 1.

YPH149 and YPH152 grown in minimal medium (selecting URA^+) show a more intense 90-kb chromosome fragment.

REFERENCE

Carle, G.F. and M.V. Olson. 1985. An electrophoretic karyotype for yeast. *Proc. Natl. Acad. Sci.* **82:** 3756–3760.

Appendix F

Strains

Experiment I		**Looking at Yeast Cells**
1-1	TSY623	*MAT*α *ade2 his3 leu2 ura3*
1-2	TSY807	*MAT***a** *his3 leu2 lys2 ura3*
1-3	TSY800	*MAT***a**/α *ADE2/ade2 his3/his3 leu2/leu2 lys2/LYS2 ura3/ura3*
1-4	TSY481	*MAT***a**/α *ADE2/ade2 his3/his3 leu2/leu2 lys2/lys2 ura3/ura3* [pTS568]
1-5	TSY514	*MAT***a** *his3 leu2 lys2 ura3* [pTS592]

Experiment II		**Isolation and Characterization of Auxotrophic, Temperature-sensitive, and UV-Sensitive Mutants**
2-1	S288C	*MAT*α *mal gal2*
2-2	D665-1A	*MAT***a**

Experiment III		**Meiotic Mapping**
3-1	1385-3B	*MAT***a** *his4-G met13 lys2 ho::LYS2*
3-2	1384-10D	*MAT*α *his4-N met4 ura3 leu2 trp1 lys2 ho::LYS2*
3-3	1385-3B x 1384-10D	
3-4	FW786	*MAT***a** *ade8*
3-5	FW787	*MAT*α *ade8*
3-10	ERX-17C	*MAT***a** *his4-N met4 ade8*
3-11	ERX-3C	*MAT*α *his4-N met4 ade8*
3-12	ER3**a**	*MAT***a** *his4-G met13 ade8*
3-13	ER3α	*MAT*α *his4-G met13 ade8*

Experiment IV		**Mitotic Recombination and Random Spore Analysis**
4-1	TSY812	*MAT*α *can1 hom3 leu2 lys2 ura3*
4-2	TSY813	*MAT***a** *ade2 his1 lys2 trp1*

EXPERIMENT V		Transformation of Yeast
5-1	TSY623	*MAT*α *ade2-101 his3-Δ200 leu2-3,112 ura3-52*
5-2	TSY502	*MAT***a** *his3-Δ200 leu2-3,112 lys2-801 ura3-52 tub4-34*
5-3	TSY807	*MAT***a** *his3-Δ200 leu2-3,112 lys2-801 ura3-52*
5-4	TSY808	*MAT***a** *ade2-101*

EXPERIMENT VI		Cytoduction and Karyogamy
6-1	DG21	*MAT***a** *kar1 aro2 ade3 ura3*
6-2	DDM30B	*MAT*α *ade2 can1 ura3Δ trp1Δ leu2 ade5 cyh2*^{R} *lys5*
6-3	PT1	*MAT***a** *ile hom3 can*^{R}
6-4	PT2	*MAT*α *ile hom3 can*^{R}

EXPERIMENT VII		Gene Replacement
7-1	BY4732	*MAT*α *his3Δ200 met15Δ0 trp1Δ63 ura3Δ0*
7-2	UCC1118	*MAT*α *ade2Δ::hisG his3Δ200 leu2Δ0 lys2Δ0 met15Δ0 trp1Δ63 ura3Δ0 Δhhf2-hht2::MET15 Δhhf1-hht1::LEU2* [pRM200U *(URA3 CEN ARS HHF2-HHT2)*]

EXPERIMENT VIII		Isolation of *ras2* Suppressors
8-6	1784TRP	*MAT*α *RAS2 his4 ura3 leu2 can1*
8-7	AMP141	*MAT***a** *ras2-530::LEU2 his4 ura3 leu2 trp1 can1*
8-8	AMP142	*MAT*α *ras2-530::LEU2 his4 ura3 leu2 lys2 can1*

EXPERIMENT IX		Manipulating Cell Types
9-1	CKY8	*MAT*α *ura3-52 leu2-3,112*
9-2	AAY1017	*MAT*α *his1*
9-3	AAY1018	*MAT***a** *his1*
9-4	YSC006	*MAT*α *ura3 ade2-1 trp1-1 can1-100 leu2-3,112 his3-11,15[psi*$^+$*]GAL*$^+$
9-5	YSC005	*MAT***a** *ura3 ade2-1 trp1-1 can1-100 leu2-3,112 his3-11,15[psi*$^+$*]GAL*$^+$

EXPERIMENT X **Isolation of Suppressors of Telomeric Silencing by Insertional Shuttle Mutagenesis**

10-1	UCC3505	*MATα ura3-52 lys2-801 ade2-101 leu2-Δ1 his3-Δ200 Δppr1::LYS2 adh4::URA3-TEL-VIIL ADE2-TEL-VR*

EXPERIMENT XI **Two-hybrid Protein Interaction Method**

11-1	Y190	*MAT***a** *gal4 gal80 his3-Δ200 trp1-901 ade2-101 leu2-3,112 lys2:GAL-HIS3:LYS2 ura3-52:GAL-lacZ:URA3* Cyh^r
11-2	Y187	*MATα gal4 gal80 his3-Δ200 trp1-901 ade2-101 ura3-52 leu2-3,112 ura3-52:GAL-lacZ:URA3*
11-3	TSY801	Y190[pTS434]
11-4	TSY802	Y187[pTS721]
11-5	TSY803	Y190[pAS1]
11-6	TSY804	Y187[pACTII]
11-7	TSY805	Y190[pTS428]
11-8	TSY806	Y190[pTS430]

Appendix G

Counting Yeast Cells with a Standard Hemocytometer Chamber

(A. Kistler and S. Michaelis)

Using a **10x** microscope objective (and a 10x ocular), the ''large'' square shown in the circle below will fill your field of view. This square traps a volume of 0.1 µl. Therefore, to calculate the number of cells/ml in a particular culture, use the following formula:

cells/square x 10^4 x dilution factor = # cells/ml in your culture

STANDARD HEMOCYTOMETER CHAMBER

(Modified with permission of Sigma-Aldrich Co. [copyright 1994])

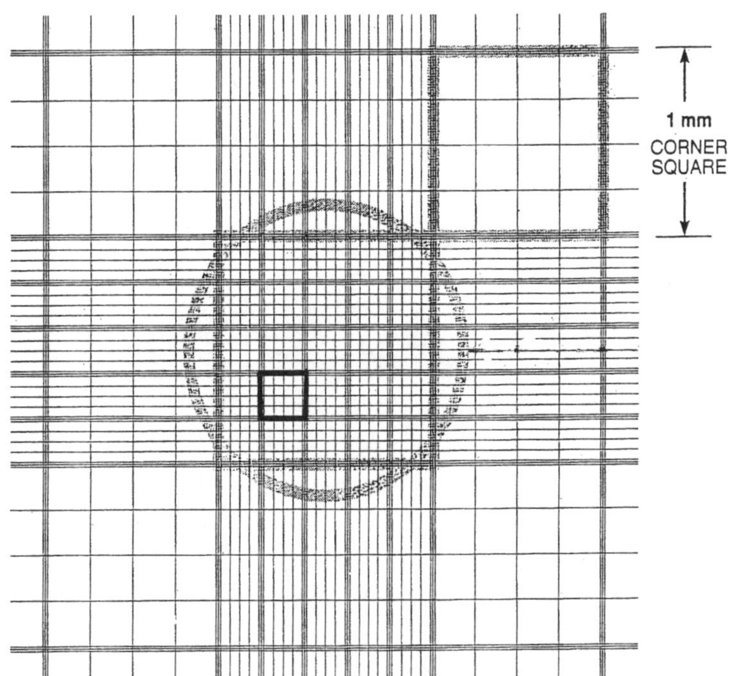

A general guide to typical cell counts for yeast cultures:

Culture	OD_{600}	Cell count	Dilution to count with hemocytometer
Sat'd YEPD	25	3 x 10^8/ml	10^{-2} or 10^{-3}
Log YPD	0.25	3 x 10^6/ml	10^0 or 10^{-1}
Sat'd SC-LYS	5	1 x 10^8/ml	10^{-1} or 10^{-2}

Generally, it is optimal to use a **40x** microscope objective, in which case only a portion of the circled field is visible, i.e., several ''small'' squares (a small square is indicated at lower left above). Then, count the cells in 5 small squares and use the following formula:

cells in 5 ''small'' squares x 5 x 10^4 x dilution factor = # cells/ml in your culture